Product Testing:
The Chemistry of Ice Cream

PACT Chemical Technology Resources

Product Testing:
The Chemistry of Ice Cream

by Dianne N. Epp

Series Editor
Mickey Sarquis, Director
Center for Chemistry Education

Terrific Science Press
Miami University Middletown
Middletown, Ohio

Terrific Science Press
Miami University Middletown
4200 East University Boulevard
Middletown, OH 45042
513/727-3269
cce@muohio.edu

This monograph is intended for use by teachers, chemists, and properly supervised students. Users must follow procedures for the safe handling, use, and disposal of chemicals in accordance with local, state, federal, and institutional requirements. The cautions, warnings, and safety reminders associated with the doing of experiments and activities involving the use of chemicals and equipment contained in this publication have been compiled from sources believed to be reliable and to represent the best opinions on the subject as of the date of publication. Federal, state, local, or institutional standards, codes, and regulations should be followed and supercede information found in this monograph or its references. The user should check existing regulations as they are updated. No warranty, guarantee, or representation is made by the author or by Terrific Science Press as to the correctness or sufficiency of any information herein. Neither the author nor the publisher assumes any responsibility or liability for the use of the information herein, nor can it be assumed that all necessary warnings and precautionary measures are contained in this publication. Other or additional information or measures may be required or desirable because of particular or exceptional conditions or circumstances, or because of new or changed legislation.

ISBN: 1-883822-25-4

This material is based upon work supported by the National Science Foundation under Grant Number DUE 9950011. Any opinions, findings, and conclusions or recommendations expressed in this material are those of the author and do not necessarily reflect the views of the National Science Foundation.

● Table of Contents

● Acknowledgments

The author and editor wish to thank the following individuals who have contributed to the development of *Product Testing: The Chemistry of Ice Cream.*

Terrific Science Press Design and Production Team

Document Production Manager: Amy Hudepohl
Technical Editing: Amy Hudepohl, Kate McCann, Lisa Taylor
Production: Dawnetta Chapman, Dot Lyon, Kate McCann, Tom Schaffner
Illustrations: Carole Katz
Cover Design and Layout: Susan Gertz

Content Specialists, Reviewers, and Classroom Testers

William Bleam, Jr., Radnor High School, Radnor, PA
Frank Cardulla, Libertyville High School (retired), Libertyville, IL
James V. Chambers, Professor Emeritus of Food Science, Purdue University,
    West Lafayette, IN
Robert T. Marshall, Professor Emeritus, Department of Food Science, University
    of Missouri-Columbia, Columbia, MO
M. Larry Peck, Professor and Director of First Year Chemistry Programs,
    Chemistry Department, Texas A&M University, College Station, TX
J. Timothy Perry, Mt. Hebron High School, Ellicott City, MD
George Rizzi, The Procter & Gamble Company (retired), Cincinnati, OH
Jim Schwarz, Kroger Springdale Ice Cream & Beverage, Cincinnati, OH
Amy Stander, Assistant Director, PACT

● Dedication from the Author

To ice cream lovers of all ages, particularly my husband, with whom I have shared dishes of ice cream in many places; my father, who still makes his own ice cream; and my grandchildren, current and future, who represent the next generation of ice cream devotees.

● Foreword

Product Testing: The Chemistry of Ice Cream is part of the PACT Chemical Technology Resources series, which is a product of the National Science Foundation–funded Partnership for the Advancement of Chemical Technology (PACT) program at Miami University Middletown in Ohio. PACT is an industrial/academic collaborative committed to creating a well-educated, chemistry-based technical workforce. Members of the PACT Consortium share the goal of bringing chemistry and chemical technology education into closer alignment with the skills, methods, problem solving, and content used in today's industrial and government laboratories.

Author Dianne Epp, a teacher at East High School in Lincoln, Nebraska, developed *Product Testing: The Chemistry of Ice Cream* with the specific goal of enabling high school chemistry teachers to introduce their students to one of the important aspects of chemical engineering and technology: product testing. Dianne has more than 20 years of teaching experience. She is a highly respected author of resource books for high school chemistry teachers.

In this volume, the science and production of ice cream are used as tools to illustrate the concept of product testing because ice cream is a familiar and interesting topic to students. To provide the foundation for the complex questions of product testing, this book includes background on the chemistry and history of ice cream, along with a series of activities in which students explore the ingredients, melting behavior, texture, and structure of ice cream.

A suggested companion to this monograph is the *Chemistry in Industry: Volume 3* CD-ROM also produced by PACT. One of the interactive units features the ice cream production facility of the Kroger Springdale Ice Cream & Beverage company.

We hope you find this monograph to be a useful and exciting tool to involve your students in workplace-based applications of product testing. You can find out more about the PACT Chemical Technology Resources series by accessing our website at *http://www.terrificscience.org/PACT/*.

Mickey Sarquis, Series Editor
Director, Center for Chemistry Education
Professor, Chemistry and Biochemistry
July 2001

● Preface

This monograph deals with one of the basic areas of industrial chemistry, testing for quality control of both raw materials and finished product. Although the chemical engineer may bear overall responsibility for the industrial process, responsibility for most of the actual testing lies with the laboratory technicians and other food scientists, who become key players in determining the final quality of a product by their conscientious quality-control testing. This monograph focuses on both ice cream production and the role testing plays in the production of a safe and delicious product.

This monograph is divided into six major sections: an Introduction to the Ice Cream Industry (content review for teachers) and five content sections for students, which include background readings, overheads, and laboratory activities. The laboratory activities are designed to be done in pairs by students within a 50-minute laboratory period, which is a typical format for most high schools. The last of the five student sections is a series of take-home research activities in which a problem is posed and students design their own testing protocols to investigate the problem.

Teaching Strategies and Relationship to National Science Education Standards

Section 1, "I Scream, You Scream, We All Scream for Ice Cream," sets the historical context with a discussion of the origin of ice cream and a timeline of developments in the ice cream industry. (The Cross-Curricular Integration ideas in this preface include history extensions.) Section 1 reinforces the National Research Council (NRC) Science Education Content Standard G concerning the history and nature of science, helping students see science as a human endeavor.

Section 2, "The Ingredients: Their Properties and Roles in Ice Cream Formulation," introduces students to the chemistry of milk, the major raw ingredient for ice cream, and to the terminology used in ice cream manufacturing. The laboratory activities in this section parallel the tests laboratory technicians perform on incoming milk samples, such as titratable acidity and total solids. This section addresses NRC Content Standard B concerning physical science, particularly the structure and properties of matter.

In **Section 3,** "The Production Process: From Blending to Freezing," students "tour" an ice cream plant and carry out a laboratory activity to determine the depression of freezing point as a function of various ice cream sweeteners. Understanding the technology of ice cream production reinforces NRC Content Standard E concerning science and technology, particularly the part dealing with how technological processes are used to solve problems.

In **Section 4,** "Quality Assurance: The Role of the Laboratory Technician in Product Testing," laboratory activities similar to those carried out by laboratory technicians during quality control support NRC Content Standard G concerning career information. The NRC life science Content Standard C is addressed by microbial testing, which involves studying the behavior of microorganisms.

Section 5, "Ice Cream Take-Out," includes research activities that address NRC Content Standard A concerning science as inquiry by requiring students to design research activities that allow them to evaluate a product, implement the experimental activities, collect and interpret data, and communicate their results.

Many communities have local ice cream manufacturing plants or dairies where the information in this monograph is applied in daily practice. As a culmination to this study, a field trip to one of these industries would allow students to observe a real-life application of the techniques they have been studying.

Cross-Curricular Integration
The following topics suggest ways to integrate a scientific study of ice cream production into other areas of your curriculum.

Biology/Botany

- Vanilla is a flavoring extracted from a variety of plants. Have students study these plants and their taxonomic relationships.

Consumer Science

- Have students compare their perceptions of ice cream before and after they perform the activities in this book.

History

- Have students study the effects of Prohibition and World War II on the ice cream industry.

Language Arts

- Have students collect literary references to ice cream, including metaphors, quotes, or situations in stories or poems where ice cream plays a role.

- Have students explore the unique "soda jerk slang," including terms such as jimmies, belch water, cow juice, and mud.

Nutrition

- Have students assess the nutritional value of various ice cream products and study their place in a balanced diet.

Marketing/Economics

- Have students design packaging and devise advertising campaigns for a variety of ice cream products.

- Have students research the economic impact of the ice cream industry on the global market.

- Study the economic effect of the change in nutritional labeling laws resulting in ice milk being relabeled as "light ice cream."

● Safety Procedures

Experiments, demonstrations, and hands-on activities add relevance, fun, and excitement to science education at any level. However, even the simplest activity can become dangerous when the proper safety precautions are ignored or when the activity is done incorrectly or performed by students without proper supervision. While the activities in this book include cautions, warnings, and safety reminders from sources believed to be reliable, and while the text has been extensively reviewed, it is your responsibility to develop and follow procedures for the safe execution of any activity you choose to do. You are also responsible for the safe handling, use, and disposal of chemicals in accordance with local and state regulations and requirements.

Safety First

- Collect and read the Materials Safety Data Sheets (MSDS) for all of the chemicals used in your experiments. MSDSs provide physical property data, toxicity information, and handling and disposal specifications for chemicals. They can be obtained upon request from manufacturers and distributors of these chemicals. In fact, MSDSs are often shipped with the chemicals when they are ordered. These should be collected and made available to students, faculty, or parents for information about specific chemicals used in these activities.

- Read and follow the American Chemical Society Minimum Safety Guidelines for Chemical Demonstrations on the next page. Remember that you are a role model for your students—your attention to safety will help them develop good safety habits while assuring that everyone has fun with these activities.

- Read each activity carefully and observe all safety precautions and disposal procedures. Determine and follow all local and state regulations and requirements.

- Never attempt an activity if you are unfamiliar or uncomfortable with the procedures or materials involved. Consult a college or industrial chemist for advice or ask him or her to perform the activity for your class. These people are often delighted to help.

- Always practice activities yourself before using them with your class. This is the only way to become thoroughly familiar with an activity, and familiarity will help prevent potentially hazardous (or merely embarrassing) mishaps. In addition, you may find variations that will make the activity more meaningful to your students.

- You, your assistants, and any students participating in preparation or performance of an activity must wear appropriate personal protective equipment.

- Special safety instructions are not given for everyday classroom materials being used in a typical manner. Use common sense when working with hot, sharp, or breakable objects. Keep tables or desks covered to avoid stains. Keep spills cleaned up to avoid falls.

American Chemical Society Minimum Safety Guidelines for Chemical Demonstrations

This section outlines safety procedures that Chemical Demonstrators must follow at all times.

❶ Know the properties of the chemicals and the chemical reactions involved in all demonstrations presented.

❷ Comply with all local rules and regulations.

❸ Wear appropriate eye protection for all chemical demonstrations.

❹ Warn the members of the audience to cover their ears whenever a loud noise is anticipated.

❺ Plan the demonstration so that harmful quantities of noxious gases (e.g., NO_2, SO_2, H_2S) do not enter the local air supply.

❻ Provide safety shield protection wherever there is the slightest possibility that a container, its fragments, or its contents could be propelled with sufficient force to cause personal injury.

❼ Arrange to have a fire extinguisher at hand whenever the slightest possibility for fire exists.

❽ Do not taste or encourage spectators to taste any nonfood substance.

❾ Never use demonstrations in which parts of the human body are placed in danger (such as placing dry ice in the mouth or dipping hands into liquid nitrogen).

❿ Do not use "open" containers of volatile, toxic substances (e.g., benzene, CCl_4, CS_2, formaldehyde) without adequate ventilation as provided by fume hoods.

⓫ Provide written procedure, hazard, and disposal information for each demonstration whenever the audience is encouraged to repeat the demonstration.

⑫ Arrange for appropriate waste containers for and subsequent disposal of materials harmful to the environment.

Teacher Background

The Importance of Product Testing

How are the safety and quality of ice cream ensured as raw materials are processed into the final product? This is a critical question that must be answered by the ice cream industry, and it is the underlying reason for quality-control testing.

In both batch plants and continuous-process plants, basic manufacturing steps include composing or blending the ice cream mix, homogenizing and pasteurizing, cooling, aging (in a storage tank), freezing, packaging, hardening, storing, and loading out the finished product. Laboratory technicians and other food scientists ensure the quality of the ice cream by conducting quality-control tests at each step of the process. Figure 1 illustrates the sequence of these steps, which are detailed throughout this book.

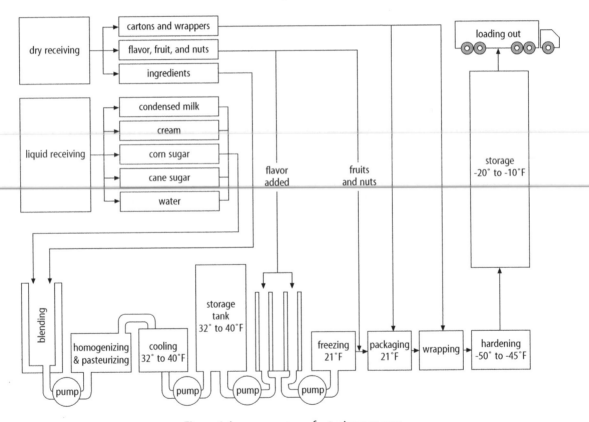

Figure 1: Ice cream manufacturing process

The selection of high-quality raw ingredients is critical to the quality of the finished product. In compliance with state regulations, a company's corporate office sets the specifications for raw materials. The ice cream plant technicians then ensure that raw materials meet company and government specifications. Tests on raw materials may be conducted periodically for materials such as stabilizers and emulsifiers, which are purchased from suppliers who provide quality certification. For materials such as fresh dairy products, tests may be performed on each batch received. Fresh, sanitary dairy products are the key ingredients in high-quality ice cream, so technicians sample and rigorously test all incoming dairy products. Tests are performed for temperature, flavor and aroma, specific gravity, percentage of milkfat and total solids, numbers of bacteria, presence of antibiotics, and titratable acidity. After the raw ingredients have been blended into the final ice cream mix, compositional tests are run on the blended mix.

During the processing of the mix into the final product, technicians check for completeness of pasteurization and homogenization. As the saying goes, "The proof of the pudding is in the eating," so a taste test and a visual inspection of the finished product are critical parts of the process. In addition, the finished ice cream is subjected to a battery of other tests, including final microbial analysis, milkfat content, meltdown and shape retention tests, and microscopic examination of texture. Ice creams containing fruits, nuts, swirls, and other mix-ins are examined for proper distribution of these materials. In the hands of competent laboratory technicians and conscientious plant workers, the high quality and safety of the product are ensured.

The Ingredients: Their Properties and Roles in Ice Cream Formulation

Early ice cream's composition included a relatively limited number of ingredients—cream, milk, sugar, and sometimes eggs or gelatin as a stabilizer. Gradually other ingredients, such as condensed milk, nonfat dry milk, whey powder, and butter, came into wide use as part of the ice cream mix. Today, though a wide variety of individual ingredients is used in each category, the basic categories of ice cream ingredients are milkfat, nonfat milk solids (NMSs), sweeteners, stabilizers and emulsifiers, flavorings, water, and air.

Milkfat

Of all the ingredients in ice cream, milkfat content varies the most widely—from 10–16%, depending on governmental standards of identity, quality, price, and competition. According to the U.S. Department of Agriculture (USDA) labeling standards, a minimum of 10% milkfat must be present for the product to be labeled "ice cream" without a modifier such as "light."

Milkfat is produced only in the mammary gland of mammals and occurs in milk as tiny fat globules held in suspension as an emulsion. On average, milk contains about 2.5 billion fat globules per milliliter ranging in size from 0.8–2.5 micrometers (μm) in diameter. (See Figure 2.) These globules have a surface layer of protein and lecithin known as the fat globule membrane.

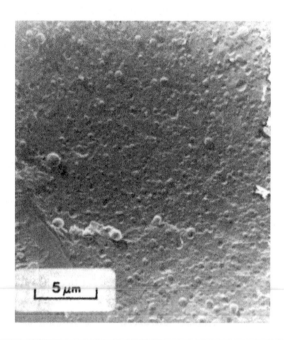

Figure 2: Electron micrograph of fat globules in ice cream mix

The tendency of fat globules to cluster together gives body to ice cream. Body refers to the quality of firmness and to the melting behavior of the ice cream, and it is often characterized by the way the ice cream "feels" in the mouth. Milkfat not only supplies richness of taste but also carries fat-soluble flavors such as vanilla to the tongue. These flavors are then slowly released as the fat melts so that they remain intense over a significant period of time. Ice cream texture is also directly related to the milkfat content—the higher the milkfat content, the less crystalline the ice cream will appear to the consumer.

Milkfat is composed of more than 400 different fatty acids that are attached to glycerol to form triglycerides. (See Figure 3.) However, 10 fatty acids predominate. The fatty acids are attached to the glycerol in combinations (mixed triglycerides) that keep the melting point of the fat below the body temperature of the cow. Therefore, no crystalline fat forms within the animal. Since short-chain fatty acids, such as 4-carbon butyric acid, and unsaturated ones, such as 18-carbon oleic, have relatively low melting points, they are attached to glycerol molecules to which are attached the higher-melting 14-, 16-, and 18-carbon saturated fatty acids. An unsaturated fatty acid contains one to four carbon-to-carbon double bonds. Most of them have 18–24 carbons within the chain. Saturated fatty acids contain only single carbon-to-carbon bonds. Simple triglycerides, formed from glycerol and three molecules of the same fatty acid, rarely occur in milk.

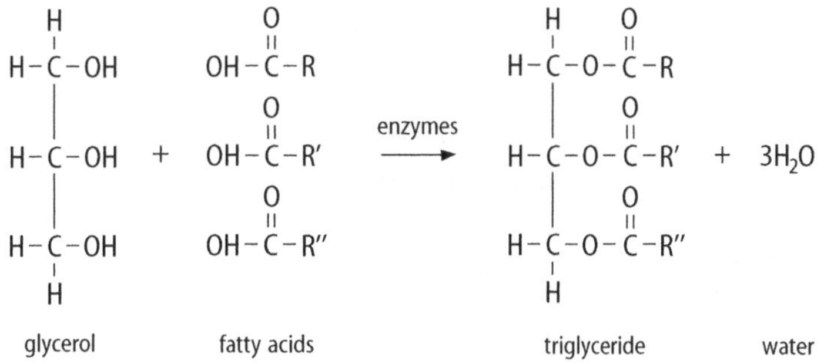

Figure 3: Formation of a mixed triglyceride
(R, R', and R" represent long carbon chains)

Milkfat is associated with small amounts of phospholipids, in which a portion of the fatty acids is replaced with phosphoric acid. When this phosphoric acid forms an ester with the nitrogen base choline, part of the vitamin B complex, a lecithin molecule is formed. Milkfat contains about 0.6% lecithin. The general structure of a lecithin molecule is shown in Figure 4.

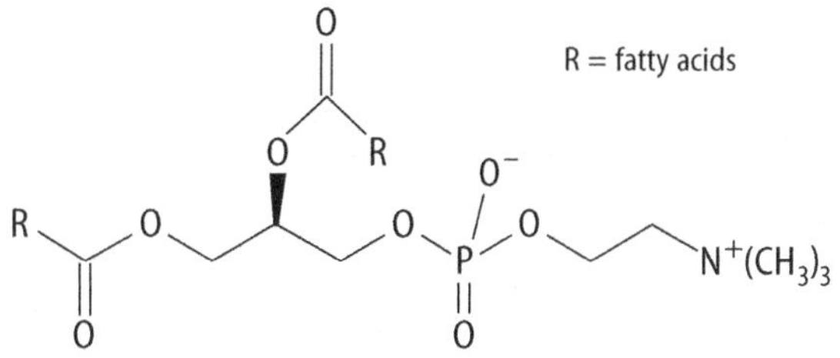

Figure 4: General structure of a lecithin molecule

Other minor components associated with milkfat include the phospholipid cephalin; the sterols cholesterol and ergosterol; the carotenoids carotene and xanthophyll; and the vitamins A, D, and E.

Nonfat Milk Solids

If the milkfat is removed from milk, as in skim milk, the remaining solids are termed nonfat milk solids (NMSs). These solids consist of approximately

37% protein, 55% lactose (milk sugar), and 8% minerals. The protein portion of NMSs is made up of 80% casein and 20% whey protein. The major proteins in casein are alpha, beta, and kappa, which are found in combination with calcium and phosphorus as calcium phosphocaseinate. Whey proteins include immune globulin, alpha lactalbumin, beta lactoglobulin, and serum albumin. The proteins in NMSs improve the body and texture of ice cream by making it more compact and smooth.

Lactose, a disaccharide of galactose and glucose linked together, adds slightly to the sweetness of ice cream. If the proportion of NMS is too high in the whey solids of the ice cream mix, some of the lactose may crystallize during storage, causing the texture of the ice cream to feel "sandy." The structure of lactose is shown in Figure 5.

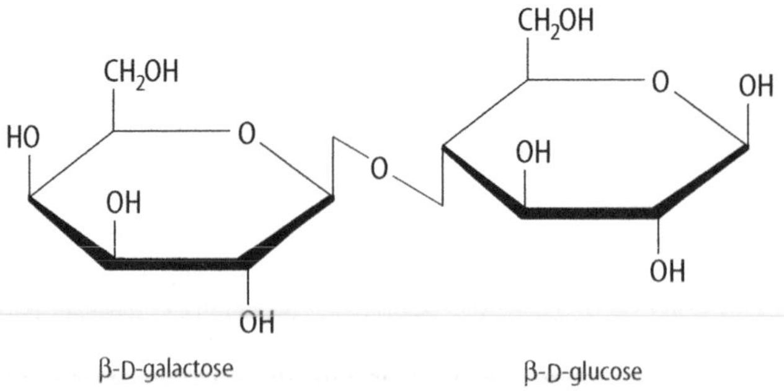

β-D-galactose β-D-glucose

Figure 5: Structure of lactose

Mineral salts of calcium, sodium, potassium, magnesium, iron, copper, and other trace elements occur in ice cream in the form of citrates, phosphates, chlorides, or oxides and provide nutritional value and the slightly salty taste that contributes to the final flavor of ice cream.

Sweeteners

Many kinds of sweeteners, including cane and beet sugars, corn sweeteners, maple and brown sugars, molasses, and honey are found in ice cream. Sucrose may be used as the only sweetener in some ice cream mixes, but frequently corn syrup is also included because it is less expensive. A combination of 70% sucrose with 30% corn syrup is often used in an ice cream mix.

The structures of sucrose (obtained from cane and beet sugars) and dextrose (the major sweetening agent in corn sweeteners) are shown in Figure 6.

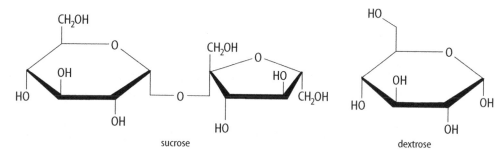

Figure 6: Structures of sucrose and dextrose

The total amount of sweetener in ice cream might vary from 12–20%, with 14–16% of sucrose providing the optimal desired sweetness. Sugars, because they are dissolved solids in an ice cream mix, affect the freezing point of the solution. This freezing point can be calculated from the concentration and molecular weight of the sugar. A 2% increase in sucrose content lowers the freezing point approximately 0.23°F (0.13°C). (Note that using the Fahrenheit scale is the industry standard, but Celsius equivalents are included in this book for pedagogical reasons.) For given weights and volumes of solvent, the effect on the freezing point is inversely proportional to the molecular weight of the sugar. Corn sugar (a monosaccharide known as both dextrose and D-glucose) is approximately half as massive as sucrose (a disaccharide). This means that corn sugar will lower the freezing point of ice cream approximately twice as much as the same mass of sucrose, so an ice cream mix containing corn sugar will tend to melt quickly due to its lower melting point. Because individuals perceive sweetness differently (and because no chemical test is available to judge sweetness), the common practice is to compare the relative sweetness of other sugars to sucrose, with sucrose being assigned a value of 100.

Sweeteners made by hydrolyzing corn starch vary in sweetness relative to sucrose from about 23 for the 20% hydrolyzed form of corn syrup to 74 for the fully hydrolyzed form called corn sugar. Corn syrups range in relative sweetness from 23–72 and degree of hydrolysis from 20–68%, respectively. As corn starch is increasingly hydrolyzed, it increases in sweetness and in its effect on freezing point. Brown sugar, maple sugar, and honey are infrequently used as sweeteners and are primarily included when their particular flavors are desired.

Stabilizers

Stabilizers provide ice cream with smoothness of texture, firmness of body, and slowness of melting, which helps retard ice crystal growth during storage. The stabilizers function either by creating a gel structure or by taking the water into themselves as water of hydration. The amount of stabilizers used in ice cream varies between 0.1 and 0.5%. Although gelatin was the first commercial stabilizer used and is still in use, numerous other stabilizers, such as agar, various gums, propylene glycol alginate, sodium alginate, carrageenan, and sodium carboxymethyl cellulose (CMC) have become increasingly important. Many stabilizers are derived from natural products, such as gelatin, an animal protein, and carrageenan, a seaweed extract. Some stabilizers are chemically modified natural products, such as CMC and propylene glycol alginate, both of which are polysaccharides obtained from botanical sources.

Emulsifiers

Emulsifiers hold together two immiscible liquids. The emulsifier concentrates at the interface between fat and water, reducing the overall surface tension, improving the whipping quality of the mix, and producing a smoother ice cream that flows more easily when drawn out of the freezer. While milk proteins, lecithin, phosphates, and citrates act as natural emulsifying agents in milk, several commercial agents have been shown to be more effective for use in ice cream. Commercial emulsifying agents used in making ice cream include mono- and diglycerides and derivatives of hexahydric alcohols, glycol, and glycol esters. Figure 7 illustrates how an emulsifier holds fat globules in suspension following homogenization.

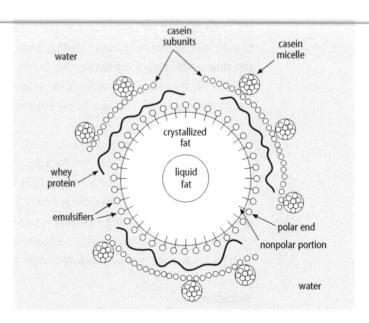

Figure 7: Cross-section representation showing an emulsifier in a suspension of globules with the polar end of the emulsifying agent attracted to the water and the nonpolar end attracted to the fat globule

Flavoring

It may not be surprising to learn that, even though some manufacturers have recipes for over 500 flavors, vanilla and chocolate still dominate the ice cream flavor market! Flavors are classified by their type and intensity. Customer satisfaction is directly related to a "well-flavored" product. In general, the flavor should be only intense enough to be easily identified and to provide a delicate, pleasing taste.

Federal standards for frozen desserts require adding a modifier to a flavor's name when artificial flavoring is used with or in place of the natural flavoring. For example, when only pure vanilla is used, ice cream can be labeled "vanilla ice cream." However, if pure vanilla dominates but artificial vanilla flavoring is added, the product must be labeled "vanilla flavored ice cream." When artificial flavoring dominates over the pure vanilla or is the only flavoring used, the product must be labeled "artificially flavored vanilla ice cream." The rule holds for other flavors as well.

Water and Air

Two essential but often overlooked components of ice cream are water and air. Water exists in both liquid form and solid form. The liquid portion of ice cream contains dissolved substances containing suspended ice crystals, air cells, fat globules, casein micelles, and stabilizers. As the temperature is lowered the amount of liquid decreases.

Air cells provide softness and smoothness in the frozen product. These cells are surrounded by a thin protein layer on which fat globules are adsorbed. Some of these fat globules coalesce as others rupture, releasing liquid fat that cause the globules to stick together. This process stabilizes the air cells, imparts a dry appearance, and slows the meltdown rate of the ice cream. Figure 8 shows the physical structure of ice cream represented on a microscopic scale.

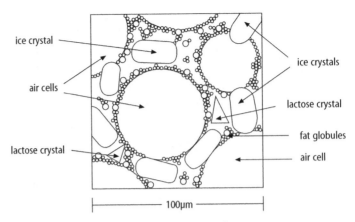

Figure 8: Microscopic representation of ice cream structure

The water in ice cream comes primarily from fluid dairy products; milk is 85–88% water. The ice cream manufacturing process incorporates air into the mix by constantly agitating the mix. This air whipped into the ice cream mix during the freezing process and the expansion of the water as it freezes contribute to overrun, an increased volume of the finished product over the volume of the original liquid mix. Maintaining a proper amount of overrun is critical to both quality and profit.

The Production Process: From Blending to Freezing

Once the ingredients are selected according to an established company formulation, they are weighed and blended together into the ice cream mix. High-speed blenders are often used to agitate the mix, thoroughly incorporating the powdered material into the liquid ingredients. When the ingredients are blended, the mixture is ready for homogenizing, pasteurizing, cooling, aging, and freezing.

Homogenizing

The ice cream mix is homogenized by being forced through a small opening at high pressure. The opening consists of a valve and seat with parallel adjacent surfaces surrounded by an impact ring against which the mix impinges as it leaves the valve. The forces set up by this system break up the fat globules to less than 2 μm in diameter. This creates a permanent and uniform suspension in which the fat will not rise and form a cream layer. Homogenization also improves whipping ability, slows the "churning" of fat during freezing, improves the smoothness of the ice cream, and increases resistance to melting.

Pasteurizing and Cooling

Pasteurizing the ice cream mix destroys all pathogenic bacteria and organisms that cause spoilage. During pasteurization, hydration of proteins and stabilizers also takes place. Batch pasteurization and continuous high-temperature short-time (HTST) methods may be used. Batch pasteurizers are vats equipped with steam jackets or hot-water jackets. The mix is placed in the vat, heated to at least 155°F (68°C) for 30 minutes, then rapidly cooled to below 40°F (4°C).

In a continuous HTST method, some preheating is used to increase the solubility of various components. Then the mix is run through a pasteurization/homogenization section, followed by a cooling section. Large ice cream plants use continuous systems in which heat is transferred from the hot mix to the incoming cool mix. This "recycling" of heat, called regeneration, makes a continuous process more economical than a batch method.

Aging

Following pasteurizing and cooling, the mix is aged for about 4 hours and sometimes overnight. This process allows time for the fat globules to cool down and resolidify and for the proteins and polysaccharides to become fully hydrated. Aging takes place in refrigerated storage tanks held at the lowest temperature possible without freezing (at or below 39°F [4°C]). If called for in the formula, liquid flavors, fruit purées, and colorants are added to the cooled mix just prior to freezing.

Freezing

The function of the critical freezing process is to rapidly freeze a portion of the water in the mix while using agitation to whip air into the mix. This process is typically performed by a continuous "barrel" freezer. Ice cream mix is metered into the back of the barrel and exits the front of the barrel in an expanded, soft-frozen form. Continuous barrel freezers vary in size from small units that can produce 20 gallons of ice cream per hour to large, multibarrel units that can produce 800 gallons per hour.

The continuous barrel freezer contains a tube-type heat exchanger enveloped by a jacket generally containing liquid ammonia as the refrigerant. (See Figure 9.)

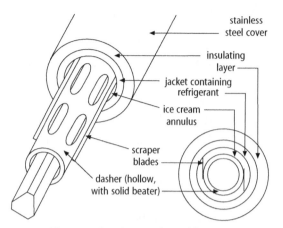

Figure 9: Continuous barrel freezer

The temperature of the ice cream mix moving through the barrel drops rapidly as heat is transferred from the mix to the liquid ammonia, causing the ammonia to vaporize. As the freezing point is reached inside the cylinder, the water in the mix begins to form ice crystals. Because these crystals are almost pure water, the amount of water (solvent) present in the unfrozen mix decreases and the dissolved solids become more concentrated, which in turn lowers the freezing point of the unfrozen mix.

For an average ice cream mix, the initial freezing point is 27.0 to 28.0°F (−2.8 to −2.2°C), which reflects primarily the freezing point depression due to the sugar content of the mix. As the temperature drops, the percentage of frozen water increases. (See Figure 10.)

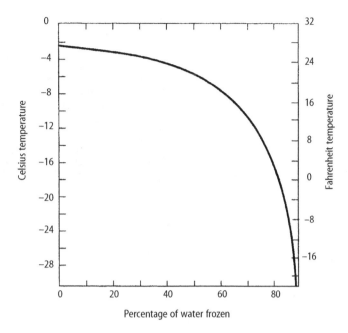

Figure 10: Typical freezing point curve for ice cream giving percentage of water frozen at various temperatures

Depending on the temperature at which the ice cream is drawn out of the freezer, anywhere from 33–67% of the water will have crystallized. During the hardening process further crystallization takes place, but since ice cream is seldom, if ever, frozen below −67°F (−55°C), some water remains in liquid form.

In a continuous barrel freezer, the mix is pumped into one end of the cylinder and drawn out the other end in a matter of 30 seconds. Fast freezing promotes a smooth product because ice crystals that form quickly are smaller than those that form more slowly. A batch freezer, by contrast, will take approximately 10 minutes to achieve the same state. Inside the barrel, rotating blades continually scrape the ice crystals off the surface of the freezer barrel, and dashers help whip air into the mix. For an average ice cream mix, when the product leaves the barrel, about 50% of its water is frozen and the ice cream has increased approximately 100% in volume. Figure 11 is a microscopic view of the internal structure of ice cream.

a: ice crystals

b: air cells

c: unfrozen material

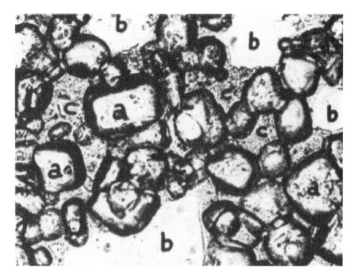

Figure 11: The internal structure of ice cream

The ice crystals average 45–55 µm, and the average size of the air cells is 100–150 µm. A thin, continuous layer of fat globules stabilizes the air cell-liquid mix interface.

As the ice cream is drawn out of the freezer, particulate matter such as nuts, candy, or fruit pieces is added. The consistency of the product at this point resembles that of soft-serve ice cream.

Finally, the ice cream is packaged and placed into a blast freezer where it is hardened at –40 to –22°F (–40 to –30°C). During the hardening process, much of the remaining water crystallizes. The ice cream then remains quite stable, with little change in quality as long as it is stored at a temperature below –20°F (–29°C).

Quality Assurance: The Role of the Laboratory Technician

Responsibility for quality assurance in the ice cream plant rests largely in the hands of laboratory technicians. Technicians not only assess the raw materials entering the plant but also test the final product. Proper laboratory testing is essential for efficient operation and for maintaining uniform quality. Tests on raw materials and products are both chemical and bacteriological in nature.

References

Arbuckle, W.S. *Ice Cream,* 4th ed.; Chapman & Hall: New York, 1986.

Arbuckle, W.S.; Marshall, R.T. *Ice Cream,* 5th ed; Chapman & Hall: New York, 1996.

Breedlove, C.H. "Vanilla," *ChemMatters.* 1988, *6* (2), 8–9.

Cook Flavoring Company Website. The Story of Vanilla. http:// www.cookflavoringco.com/ story.htm (accessed May 3, 2001).

Dickson, P. *The Great American Ice Cream Book;* Atheneum: New York, 1972.

Keeler, L.; DeVries, M.; Rupnow, J.; Wheeler, R. Department of Food Science, University of Nebraska, Lincoln, NE. Personal communication.

Liddell, C.; Weir, R. *Ices: The Definitive Guide;* Hodder and Stoughton: London, 1993.

Marshall, R.T. *Standard Methods for the Examination of Dairy Products,* 16th ed.; American Public Health Association: Washington, DC, 1993.

Mitchell, S. *Authenticity of Vanilla Samples;* Center for Chemistry Education: Middletown, OH, unpublished work, 1997.

Stogo, M. *Ice Cream and Frozen Desserts;* Wiley and Sons: New York, 1998.

• • • • • • • ## Section 1

"I Scream, You Scream,
We All Scream for Ice Cream"

● Ice Cream—A Chip Off the Old Block (Student Background 1A)

Have you ever had an "ice cream headache"? Did you know that Hippocrates, the Greek physician known as "the father of medicine," cautioned against chilling the body by ingesting cold materials?

Why should anyone run the hazard in the heat of summer of drinking iced waters which are excessively cold, and suddenly throwing the body into a different state than it was before, producing thereby many ill effects?
—Hippocrates (400 B.C.)

Chilled juices and wines were the forerunners of primitive water-based ices, which finally evolved to a variety of chilled mixtures containing milk and cream. Nero Claudius Caesar (A.D. 37–68) is said to have sent teams of runners into the mountains to relay back snow, which was flavored with fruit for the imperial table.

Another early ice cream story that may fall into the realm of myth says that, at the end of the thirteenth century, Marco Polo carried back from the Far East a recipe for a milk-based frozen dessert that laid the foundation for increasingly popular ices and sherbets. By the middle of the sixteenth century, however, clear documentation of a "food from milk sweetened with honey and frozen" appears in writings about Italian food. It is speculated that when Catherine de Médici left Italy to marry the Duke d'Orléans in 1533, the cooks and chefs who accompanied her may have introduced Italian frozen dessert to France. The earliest manuscript (circa1700) detailing the preparation of various ice creams (including apricot, rose, chocolate, and caramel, along with water-based ices) is the 84-page *L'Art de Faire des Glaces (The Making of Ice Cream)*. English references to ice cream include "crème frez," which was served at the coronation of Henry V, and regular inclusion of "crème ice" at the table of Charles I. Ice cream recipes appear in a number of English cookbooks dating back to the eighteenth century.

Ice Cream in the New World

Ice cream appears to have crossed the Atlantic with the early English colonists. Both George Washington and Thomas Jefferson are well documented as fans of the frozen dessert, but Dolly Madison glamorized it by first including it at White House state dinners.

Ice cream parlors of the nineteenth century also helped popularize ice cream as a very special delicacy. During that century, two developments made ice cream more available to the average household: 1) ice harvesting from frozen lakes and rivers (and the use of insulated icehouses) became widespread, and 2) the invention of the hand-cranked ice cream freezer in 1846 by Nancy Johnson was a great improvement over earlier "pot" methods. A few years after that invention, magazine advertisements for hand-cranked ice cream freezers became common. (See Figure 1-1.)

Buying a Peerless Freezer

WITH THE

Vacuum Screw Dasher.

Write us for information; it costs nothing and may save you money.

Our recently published booklet "**Fifty Ices**," gives full description of the Peerless, with illustrations and price list.

PEERLESS FREEZER CO., Cincinnati, O.

Figure 1-1: Mid-1800s advertisement for a hand-cranked ice cream freezer

In 1851, Jacob Fussell established the first wholesale ice cream industry in Baltimore; an associate, Perry Brazelton, successfully opened plants in St. Louis, Chicago, and Cincinnati. However, dramatic growth of the industry did not begin until the twentieth century, with the introduction of mechanical refrigeration, electric power, homogenization, and new freezers and freezing processes. By the 1920s, the value of ice cream as an essential food was generally recognized, and during the remainder of the twentieth century, the variety and popularity of ice cream and ice cream confections only increased.

Credit for the invention of the ice cream cone has been a source of ongoing historical controversy. Although an Italian immigrant, Italo Marchiony, was issued the first patent on his ice cream cone mold in 1903, he did not gain commercial success with his invention. Popularization of the ice cream cone appears to have occurred at the St. Louis World's Fair of 1904, when waferlike waffles were rolled into cornucopias and filled with ice cream. Several competing claims as to who first tried the combination can be found, but it is clear that before the fair was over, St. Louis foundries were producing baking molds for "World's Fair Cornucopias." By 1924, just 20 years after the introduction of ice cream cones, Americans were consuming an estimated 245 million of them per year.

Three unique and enduring frozen novelties were introduced between 1919 and 1924: the Eskimo Pie®, the Good Humor Bar®, and the Popsicle®. The Eskimo Pie, invented by Christian Nelson of Onawa, Iowa, was originally dubbed the "I-Scream" bar and immortalized in the Tin Pan Alley song, "I Scream, You Scream, We All Scream for Ice Cream." Nelson's partner, Russell Stover of Omaha, Nebraska, later renamed it the Eskimo Pie. Harry Burt, an ice cream parlor operator in Youngstown, Ohio, put the coated ice cream bar on a stick, calling it the "Good Humor Ice Cream Sucker."

According to legend, the Popsicle was born on a cold morning in New Jersey after Frank Epperson, who sold lemonade powder, left a glass of lemonade on a windowsill overnight. In the morning he found a frozen mass with a spoon firmly trapped in the ice. Holding the spoon "handle" and running hot water over the outside of the glass to free the frozen mass, Epperson created the first "Epsicle," later known as the Popsicle. Recognizing its potential, Epperson applied for a patent, which was issued in 1923.

During the latter half of the twentieth century, frozen desserts such as ice milk, sherbet, and frozen yogurt became increasingly popular, as shown in the graph in Figure 1-2.

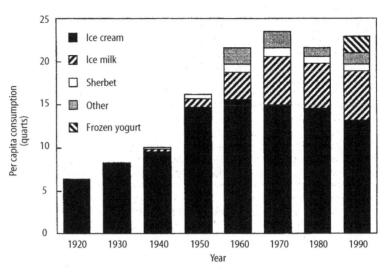

Figure 1-2: Per capita consumption of frozen desserts in the United States

Currently the United States leads the world in ice cream production and consumption. About 23–24 quarts of frozen desserts are consumed per person each year. This equals about 188 standard 4-ounce ice cream cones. Clearly, we do "all scream for ice cream"!

● Ice Cream Comes of Age (Student Background 1B)

The history of ice cream clearly predates its appearance in the New World, but the commercial production and popularization of ice cream in the United States gave rise to a giant industry that today annually produces more than enough ice cream to fill the Grand Canyon. Some of the milestones in the history of ice cream in the New World are summarized below.

1700: Ice cream is probably introduced to the New World with English colonists.

1744: A letter written by a guest of Maryland Governor Bladen describes having been served ice cream at the Governor's home.

1774: First public record of ice cream in America appears—an advertisement in a New York City newspaper by caterer Philip Lenzi announcing the availability of various confections, including ice cream.

1789: Mrs. Alexander Hamilton, wife of the Secretary of the Treasury, serves ice cream at a dinner attended by George Washington.

1811: Ice cream is served at the White House by Dolly Madison.

1848: A patent is granted to William G. Young for the first hand-cranked ice cream freezer based on an 1846 invention by Nancy Johnson.

1851: Jacob Fussell, a Baltimore milk dealer, opens an ice cream plant in Baltimore. Later he establishes plants in Washington, DC (1856) and New York City (1864).

1858: Perry Brazelton opens an ice cream plant in St. Louis.

1876: William Clewell of Reading, Pennsylvania, receives a patent for an ice cream scoop consisting of a tin and steel cone with a key release.

1879: The ice cream soda is introduced at the Centennial Exposition in Philadelphia.

1895: Pasteurizing machines are introduced.

1899: The homogenizer is invented in France. A U.S. patent is issued in 1904.

1902: The horizontal circulating brine freezer is developed.

1904: The ice cream cone appears at the St. Louis World's Fair.

1913: The continuous freezing process is patented.

1926: The counter freezer for soft-serve ice cream appears.

1935: Sherman Kelley introduces the Zeroll dipper with self-defrosting fluid in the handle.

1940s: Carry-home packages marketed through chain grocery stores gain popularity.

1942–1953: The FDA holds hearings on federal standards for ice cream.

1950s: Dairy Queen®, Tastee-Freez®, and Carvel® soft-serve drive-ins proliferate.

1961: Häagen-Dazs® markets the first superpremium ice cream.

1961: Definitions and standards for frozen desserts are approved by the FDA.

1983: Ice cream standards and regulations are revised by the FDA.

1984: July is proclaimed National Ice Cream Month.

2001: International Ice Cream Association petitions FDA to revise ice cream standards.

● Ice Cream Dates in History (Overhead 1)

1700—*L'Art de Faire des Glaces* is published in France.

1774—Philip Lenzi advertises ice cream in a New York newspaper.

1848—A patent is granted for the first hand-cranked ice cream freezer.

1851—Jacob Fussell begins the first wholesale ice cream business in Baltimore.

1876—The first ice cream scoop is patented.

1879—The ice cream soda is introduced at the Centennial Exposition, Philadelphia.

1904—The ice cream cone is popularized at the World's Fair, St. Louis.

1913—The continuous freezing process is patented.

1926—The counter freezer for soft-serve ice cream appears.

1940s—Grocery store marketing of prepackaged ice cream becomes popular.

1950s—Dairy Queen®, Tastee-Freez®, and Carvel® soft-serve drive-ins proliferate.

1961—Häagen-Dazs® markets the first superpremium ice cream.

Section 2

The Ingredients—Their Properties and Roles in Ice Cream Formulation

● The Language of Ice Cream Production (Student Background 2A)

All industries use a particular set of terms to describe important concepts specific to their technology. Here are some of the terms used in ice cream production:

Cream, heavy: Cream with a milkfat content of 36%.

Cream, light: Cream with a milkfat content of 18%.

Dasher: The movable part inside the barrel of a freezer used to stir the ice cream mix as it freezes.

Emulsifier: An ice cream additive used to stabilize the oil (milkfat) in water mixtures to form an emulsion.

Emulsion: A system (as fat in milk) consisting of one or more liquids typically dispersed as small droplets in another liquid.

Freezer, batch: Equipment used to make a single flavor and specific quantity of ice cream.

Freezer, continuous: Equipment for production of a continuous flow of partially frozen ice cream, as a continuous stream of ice cream mix is fed into the freezing cylinder.

Hardening: Process during which ice cream freezes to −15°F (−26°C) or lower following its transfer from the freezer to the hardening room.

Ice cream mix: The unfrozen prepared mixture of milk products, water, sweeteners, emulsifiers, and stabilizers.

Milkfat (butterfat): The main component of rich sweet cream. Various frozen desserts have specific characteristic percentages of milkfat.

Nonfat milk solids (NMSs): The dry solids in skim milk, which consist of 36.7% protein, 55.5% lactose (milk sugar), and 7.8% minerals, along with water-soluble vitamins and enzymes. Milk in its fluid form is 8.2% NMSs and a value of 9.0% NMSs is used in calculating ice cream formulations.

Overrun: The increase in the volume of finished ice cream over the volume of mix used. Overrun is caused by the addition of air beaten into the ice cream mix as it freezes and by the expansion of the water as it freezes.

Refrigeration: The removal of heat from a substance. In the ice cream industry, the term means cooling to temperatures between −30 and 40°F (−34 and 4°C).

Skim milk (nonfat milk, fat-free milk): Widely used to supply milk solids, this milk contains less than 0.5% fat.

Stabilizer: Substance used to produce smoothness and slow ice crystal growth during storage. Stabilizers form gel structures with water and bind the water as water of hydration.

Sweeteners: Sugars and non-nutritive substances used to give ice cream its sweet flavor. Cane or beet sugar (sucrose) alone or in combination with corn sugar is the most common sweetener. Sugar-free ice creams contain sweeteners such as aspartame along with other solids to replace the bulk (mass or solids) of the sweetener. Ice creams for diabetics commonly contain a sugar alcohol such as sorbitol. Sugar alcohols are poorly absorbed in the small intestine, so they contribute little to the blood sugar, in contrast to the high contribution by sucrose and corn sweeteners.

Total solids: Total solids are the non-water portion of the mix and determine the nutritive value, viscosity, body, and texture of the ice cream.

Whey: The semi-clear liquid drained from coagulated casein in the manufacture of cheeses. It contains most of the lactose and water-soluble proteins, minerals, and vitamins of milk. Dried whey is often added to ice cream mixes.

What Goes In Will Come Out: The Importance of Raw Materials (Student Background 2B)

High-quality ice cream requires high-quality starting ingredients. A fresh, creamy flavor can be achieved only by choosing and carefully processing fresh, pure ingredients.

Milk, the major component of ice cream, contains water, milkfat, and nonfat milk solids (NMSs). The total solids in milk account for approximately 13% of the milk by mass. These solids are dispersed in the water portion of ice cream in three ways: as a true solution (in the case of finely divided soluble solutes such as lactose), as a colloidal suspension (in the case of larger molecules such as casein), or as an emulsion (in the case of immiscible droplets such as milkfat).

Milkfat is mostly composed of triglycerides derived from glycerol and various fatty acids, as shown in Figure 2-1, where R, R', and R'' represent carbon chains of varying lengths.

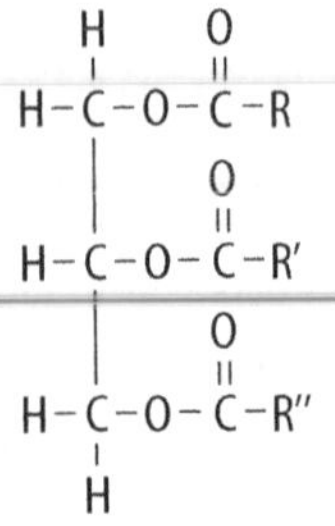

$$
\begin{array}{c}
\overset{\displaystyle H}{\underset{\displaystyle |}{|}} \quad \overset{\displaystyle O}{\overset{\displaystyle \|}{}} \\
H-C-O-C-R \\
| \qquad\quad \overset{\displaystyle O}{\overset{\displaystyle \|}{}} \\
H-C-O-C-R' \\
| \qquad\quad \overset{\displaystyle O}{\overset{\displaystyle \|}{}} \\
H-C-O-C-R'' \\
\underset{\displaystyle H}{|}
\end{array}
$$

Figure 2-1: Structure of a triglyceride

The milkfat is an emulsion of tiny globules; normal milk has about 2.5 billion fat globules per milliliter. The tendency of the fat globules to cluster together causes a cream layer to form on the surface of unhomogenized milk. Homogenization breaks the fat globules into such small droplets that a permanent emulsion forms and the fat will not rise into a cream layer.

NMSs include lactose, proteins, minerals, water-soluble vitamins, and enzymes. Casein and whey are the protein components of milk. Casein, a protein occurring only in milk, accounts for about 80% of the total milk protein. Whey contributes about 20% of the total protein of milk. The molecules of both proteins are of colloidal size, between 1 and 100 nm, with casein averaging 40–50 nm and whey particles generally ranging from 1–20 nm.

Lactose (or milk sugar, as it is commonly called), like casein, is found only in milk. Lactose is a carbohydrate disaccharide composed of two monosaccharides—glucose and galactose—linked together as shown in Figure 2-2. At warm temperatures, the bacteria usually present in milk will convert some of the lactose to lactic acid, causing the milk to sour.

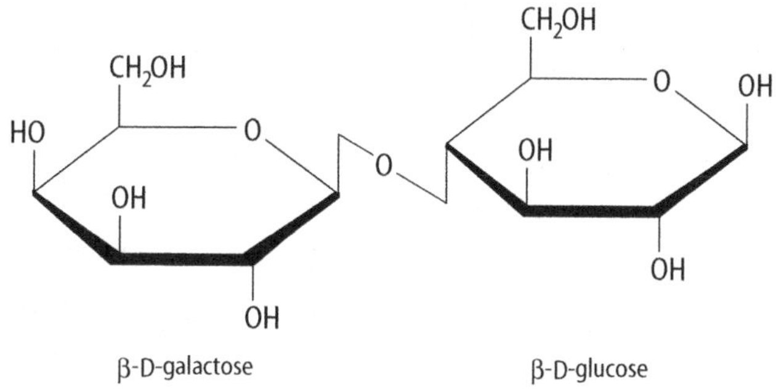

β-D-galactose   β-D-glucose

Figure 2-2: Structure of lactose

The flavor of ice cream is influenced by the amount and kind of milkfat and NMSs present in the ice cream. The full, rich, creamy flavor that we associate with premium ice cream comes mostly from the milkfat in the milk. The proteins contribute body and a smooth texture, and lactose adds to the sweetness of the ice cream even though the bulk of the sweetness comes from added sugars. Whole milk that is free from "off" odors and flavors and that has a titratable acidity below 0.18% is considered an acceptable ingredient for the manufacture of high-quality ice cream. The normal titratable acidity of milk is $0.14 \pm 0.02\%$.

Fresh milk from the cow's normal mammary gland has no developed acidity even though it has a pH of about 6.7 and requires addition of alkali to bring the pH to neutrality. The proteins, phosphates, citrates, and dissolved CO_2 in milk act as acidic buffers. Therefore, fresh milk requires about 1.4 mL of 0.1 M NaOH (since lactic acid has only one replaceable hydrogen, the molarity is equal to the normality) to raise the pH to the phenolphthalein end point of pH 8.3. This is called the apparent acidity. Acid produced by bacterial metabolism in the milk is called developed acidity.

titratable acidity = apparent acidity + developed acidity

● How Acidic Is Your Milk? (Student Activity 2A)

Because ice cream is a food, both the raw materials and the final product must meet federal Food and Drug Administration (FDA) standards. The laboratory technician is generally responsible for testing these materials. One of the common tests carried out on incoming milk and cream samples measures titratable acidity. Acidity is expressed as a percentage of lactic acid present. When whole milk is used in ice cream production, its acidity should be between 0.12 and 0.18%. Because the bacteria present in milk convert lactose to lactic acid, titratable acidity is a measure of the amount of the growth of acid-producing bacteria present in the milk. In this experiment you will analyze the titratable acidity of a milk sample to determine whether the milk sample would be acceptable to an ice cream manufacturer.

Titratable acidity is calculated by titration of the lactic acid in a milk sample with 0.10 M sodium hydroxide (NaOH). The reaction between NaOH and lactic acid is given in the following equation.

$$NaOH + HC_3H_5O_3 \rightarrow NaC_3H_5O_3 + H_2O$$

Materials

Per pair of students

- 50-mL buret and stand
- 100-mL graduated cylinder
- 50 mL 0.10 M sodium hydroxide (NaOH)
- 100-mL beaker
- 60 mL milk
- 125-mL Erlenmeyer flask
- wash bottle with distilled water
- dropper bottle with 1% alcoholic phenolphthalein indicator solution

Per class

- centigram balances

Safety and Disposal

As instructed by your teacher, follow appropriate safety procedures, including the use of personal protective equipment such as goggles and an apron. For sodium hydroxide, contact with the skin and eyes should be avoided. Should contact occur, rinse the affected area with water for 15 minutes. If the contact involves the eyes, medical attention should be sought while the rinsing is occurring. No special disposal procedures are required.

Procedure

❶ Rinse the clean buret with several 5-mL portions of the 0.100 M NaOH so that the walls of the buret are coated with a thin film of the NaOH solution. Between rinses drain the wash solution through the buret tip and discard down the drain with running water.

❷ Fill the buret with the remaining NaOH and drain out a small amount so that the buret tip is filled with the solution.

❸ Record the buret reading. This is the initial 0.100 M NaOH volume.

❹ Using a centigram balance, weigh approximately 18.00 g milk into the 125-mL Erlenmeyer flask. Record the exact mass to two decimals.

❺ Add 36 mL distilled water to the flask and swirl to mix thoroughly.

❻ Add 10 drops phenolphthalein indicator to the milk sample and mix thoroughly.

❼ Titrate with the 0.100 M NaOH to the first permanent persistence of a very light pink color in the solution.

❽ Record the buret reading. This is the final 0.100 M NaOH volume.

❾ Empty the contents of the titration flask down the drain and rinse the flask with distilled water.

❿ Repeat steps 3–9 twice more.

⓫ Calculate the percentage of acidity for each trial using the formula provided in the calculation section. Report the average titratable acidity.

Data Table for "How Acidic Is Your Milk?"			
	Trial 1	Trial 2	Trial 3
Mass of milk			
Initial mL 0.100 M NaOH			
Final mL 0.100 M NaOH			
Volume 0.100 M NaOH used in the titration			
% titratable acidity			
Average % titratable acidity			

Calculations

The titratable acidity is expressed as percentage of lactic acid present in the milk sample. Lactic acid has the formula $HC_3H_5O_3$ and a molecular weight of 90 g/mole. One mole of NaOH neutralizes 1 mole of lactic acid.

a. Calculate the number of moles of NaOH used in the titration.

b. How many moles of lactic acid were present?

c. How many grams of lactic acid were present?

d. Calculate the percentage of lactic acid in the milk sample by using the following equation.

$$\% \text{ lactic acid in milk} = \% \text{ titratable acidity} = \frac{\text{grams of lactic acid titrated}}{\text{mass of milk sample used}} \times 100$$

Questions

❶ Does your milk sample meet industry specifications for use in ice cream manufacture? Why or why not?

_____

_____

_____

❷ Lactose is a carbohydrate that can be converted to lactic acid by bacterial action. What does an unusually high titratable acidity value tell you about the milk sample?

_____

_____

_____

❸ Milk samples should be titrated as soon as possible after being taken from the incoming supply. Why will allowing the sample to stand at room temperature for an hour or more before the titration is performed affect how close the experimentally determined titratable acidity will accurately reflect the actual apparent acidity in the incoming milk sample?

_____

_____

_____

It is often said that a product is only as good as the raw materials from which it is made. That is clearly the case with ice cream. The critical job of quality assurance in the ice cream industry begins when dairy products arrive at the plant: laboratory technicians test temperature, specific gravity, percentage of milkfat, total solids, microbial and antibiotic presence, and titratable acidity of each batch entering the plant. Other raw materials such as flavorings, colorants, stabilizers, and emulsifiers that are purchased with quality certification are tested only on a periodic basis.

In Activity 2A, students measure titratable acidity, a standard test that a laboratory technician would perform on an incoming milk delivery. The test for titratable acidity is similar to the standard method for examining dairy products published by the American Public Health Association.

Safety and Disposal

It is your responsibility to review appropriate safety procedures with your students, including the use of personal protective equipment. For sodium hydroxide, contact with the skin and eyes should be avoided. Should contact occur, rinse the affected area with water for 15 minutes. If the contact involves the eyes, medical attention should be sought while the rinsing is occurring.

No special disposal procedures are required.

Procedure Notes

If students have not had prior titration experience, some instruction in titration techniques will be needed. The first persistence of light pink in the white solution may initially prove difficult for students to determine. Placing a white background under the Erlenmeyer flask will make the change more visible.

Sample Results

The following sample results were obtained by titrating 18.00 g skim milk with 0.100 M NaOH.

Sample Results for "How Acidic Is Your Milk?"			
	Trial 1	Trial 2	Trial 3
Volume of 0.100 M NaOH used in the titration	3.27 mL	3.33 mL	3.30 mL
% titratable acidity	0.164%	0.167%	0.165%

Answers to Questions

❶ *Does your milk sample meet industry specifications for use in ice cream manufacture? Why or why not?*

Answers will vary depending on results. Industry specifications for milk used in ice cream manufacture require titratable acidity to be less than 0.18%.

❷ *Lactose is a carbohydrate that can be converted to lactic acid by bacterial action. What does an unusually high titratable acidity value tell you about the milk sample?*

An unusually high titratable acidity suggests that a high amount of bacterial fermentation has occurred so that some of the lactose has been converted to lactic acid. Titratable acidity is related to the age of the sample as well as to its purity if the fermenting bacteria are present and able to grow.

❸ *Milk samples should be titrated as soon as possible after being taken from the incoming supply. Why will allowing the sample to stand at room temperature for an hour or more before the titration is performed affect how close the experimentally determined titratable acidity will accurately reflect the apparent acidity in the incoming milk sample?*

The longer a milk sample ages at a temperature favorable for bacterial growth, the more the bacteria will multiply and the more fermentation will occur. Since low temperatures inhibit bacterial growth, the samples should be kept cool until just before titration. This would minimize the developed acidity and make the experimentally determined titratable acidity closer to the apparent acidity.

● How Much of Your Milk Is Solid? (Student Activity 2B)

You will recall that milk is made up of water, milkfat, and NMSs. Everything that would be left after the removal of water is known as the total solids content. The normal range of total solids is between 11.8–15% by mass, with the mean percentage being 12.6%. When milk arrives at the ice cream plant, the laboratory technician carries out a total solids test. In this experiment you will investigate the total solids content of several milk samples.

Materials

Per pair of students
- wax pencil
- 2 porcelain evaporating dishes
- 2 Beral pipets
- 10 mL skim milk
- 10 mL whole milk
- forceps

Per class
- centigram balances
- drying oven with temperature control

Safety and Disposal

As instructed by your teacher, follow appropriate safety procedures, including the use of personal protective equipment such as goggles and an apron. No special disposal procedures are required.

Procedure
Day 1

❶ Using the wax pencil, label one evaporating dish "skim" and the second one "whole." Put your initials on the evaporating dishes so that you will be able to identify your samples when you remove them from the drying oven.

❷ Weigh and record the mass of the empty pre-dried evaporating dishes.

❸ Using a separate pipet for each sample, weigh approximately 5 g of each type of milk into its appropriately labeled evaporating dish. Record the actual mass to two decimals.

❹ Place the dishes in the drying oven, which has been set at 140°F. Dry overnight.

Day 2

❶ Remove the dishes from the drying oven using forceps and allow them to cool to room temperature. (It is best to cool dishes in a desiccator.)

❷ Weigh and record the mass of the cooled samples.

❸ Calculate the mass percentage of total solids in your two samples and post data for comparison with other class members.

❹ Calculate the average total solids for the class samples of whole and skim milk.

Data Table for Total Solids		
	Skim Milk	Whole Milk
Mass of empty dish		
Mass of filled dish		
Mass of milk sample		
Mass of dish with total solids		
Mass of total solids		
% total solids in sample		
Class average % total solids		

Questions

❶ Explain why the amount of total solids differs between the two types of milk.

_____

_____

_____

❷ Half-and-half is composed of one-half milk containing 3.3% fat and one-half light cream containing 18% fat. Predict the total solids for a sample of half-and-half (10.5% fat) and explain your reasoning.

_____

_____

_____

❸ Optional experiment: Obtain a sample of half-and-half and repeat the Procedure. Do your results match your prediction from Question 2?

_____

_____

_____

Instructor Notes for Student Activity 2B

In Activity 2B, students test for total solids, a common test that a laboratory technician would perform on an incoming milk delivery. This test is similar to the standard method for examining dairy products published by the American Public Health Association.

Safety and Disposal

It is your responsibility to review appropriate safety procedures with your students, including the use of personal protective equipment. No special disposal procedures are required.

Procedure Notes

Set the drying oven at 140°F (60°C). This will allow the samples to evaporate completely overnight without charring the remaining total solids.

Sample Results

The following results were obtained with samples of whole and skim milk.

Sample Results for "How Much of Your Milk Is Solid?"		
	Skim Milk	Whole Milk
Mass of milk sample	5.00 g	5.03 g
Mass of total solids	0.46 g	0.61 g
% total solids in sample	9.2%	12%

Answers to Questions

❶ *Explain why the amount of total solids differs between the two types of milk.*

Milk contains about 3.3% fat and 8.6% nonfat solids; skim milk contains less than 0.1% fat and about 8.7% nonfat solids.

❷ *Half-and-half is composed of one-half milk containing 3.3% fat and one-half light cream containing 18% fat. Predict the total solids for a sample of half-and-half (10.5% fat) and explain your reasoning.*

Half-and-half has a much higher milkfat content than whole milk so the total solids should be considerably higher than 14%.

❸ *Optional experiment: Obtain a sample of half-and-half and repeat the Procedure. Do your results match your prediction from Question 2?*

Sample results: A sample of half-and-half yielded a total solids value of 18%.

● Do We Really Want 31 Flavors? (Student Background 2C)

In spite of the fact that the popular "31 flavors" idea is a good marketing tool, Americans continue to prefer vanilla to all other ice cream flavors. Over 75% of all ice cream produced contains vanilla flavoring. Plain vanilla ice cream leads all others in sales, but vanilla flavoring is also used to soften the bitterness of chocolate and to enhance fruit flavors. When people taste-test an ice cream sample that contains vanilla and a sample that does not, the vanilla-flavored sample is reported to taste sweeter.

Historical references to vanilla link it to the Aztec civilization in Mexico. The fruit of the Tlilxochitl vine, now referred to as vanilla pods or beans, was used by the Aztecs to flavor a chocolate drink made with cacao beans. In 1602, Hugh Morgan, an apothecary to Queen Elizabeth I, was the first European to suggest using vanilla as a flavoring on its own. Vanilla beans are the fruit of the orchid *Vanilla planifolia.* (See Figure 2-3.)

Figure 2-3: Vanilla planifolia

The plant has been introduced to other tropical countries, notably the Bourbon Islands (Madagascar now produces 65–70% of the world's vanilla), Indonesia, and Tahiti. Tahitian vanilla originates from the plant *Vanilla tahitensis*, which produces a bean with a slightly harsher flavor.

Vanillin is the principal flavoring agent in vanilla, though more than 400 flavoring substances have been identified in extracts of the vanilla bean. The structure of vanillin is shown in Figure 2-4.

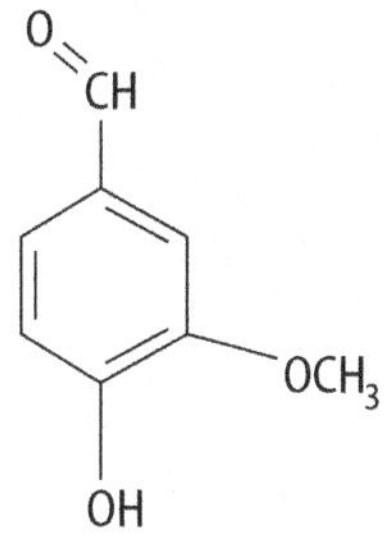

Figure 2-4: Vanillin (4-hydroxy-3-methoxybenzaldehyde)

Surprisingly, no free vanillin exists in the beans when they are harvested; it develops gradually during a period of 3–6 months following harvest, during which time the beans are alternately heated in the sun and then wrapped overnight so as to "sweat" and ferment. The cost of cured vanilla beans is very high due to the labor-intensive manner in which they are prepared. The orchid that produces the fruit requires special light, warmth, and water conditions. Flowers are pollinated by hand, the ripe vanilla beans are picked by hand, and the curing process is long and laborious.

When vanilla beans reach the manufacturer, the flavoring material must be extracted from them. Vanillin is only very slightly soluble in water but is readily soluble in the ethyl alcohol that is used to extract it from the finely cut vanilla beans. True vanilla extract has a minimum ethyl alcohol content of 35%. According to U.S. Food and Drug Administration (FDA) standards, 13.35 ounces of vanilla bean must be used to prepare 1 gallon of vanilla extract (100 g/L). Five to six ounces of this extract are needed to flavor 10 gallons of ice cream mix (about 4 mL/kg).

Vanilla flavorings are classified as pure vanilla, vanilla-vanillin blend, and imitation vanilla. Imitation vanilla is made of synthetic vanillin that is synthesized from lignin, a waste product of the wood pulp industry. Synthetic vanillin may be blended with pure vanilla to produce the vanilla-vanillin blend.

Vanilla ice cream must be labeled to reflect the flavoring used. When pure vanilla is used, the carton bears the label "Vanilla Ice Cream." The label "Vanilla-Flavored Ice Cream" indicates the use of vanilla-vanillin blend flavoring, and the label "Artificially Flavored Vanilla Ice Cream" indicates that imitation vanilla is the dominant flavoring agent.

● Vanilla (Overhead 2)

● How Pure Is Your Vanilla? (Student Activity 2C)

In the United States, the FDA regulates the production and sale of pure vanilla extract. Vanilla produced in other countries is not regulated under the same guidelines, and compounds are frequently added to make imitation vanilla taste more like pure vanilla extract. In this experiment you will use paper chromatography to establish the R_f values for four known compounds found in pure vanilla and imitation brands. Then you will test several samples of vanilla, both domestic and foreign, to determine what components are present.

Materials
Per pair of students
- 10–20 mL 50% ethyl alcohol solution
- 1-L beaker
- plastic wrap
- rubber band
- Whatman® chromatography paper rectangle (11 cm x 22 cm)
- scissors
- pencil
- ruler
- capillary tubes
- ethyl vanillin
- vanillin
- coumarin
- 4-hydroxybenzaldehyde
- 4 different commercial vanilla samples

Per class
- UV lamp set on long wavelength

Safety and Disposal
As instructed by your teacher, follow appropriate safety procedures, including the use of personal protective equipment such as goggles and an apron. No special disposal procedures are required.

Procedure
❶ Pour approximately 15 mL 50% ethyl alcohol solution into the 1-L beaker; swirl; and cover the beaker with plastic wrap, securing the wrap with a rubber band.

❷ Cut an 11-cm x 22-cm rectangle of chromatography paper. Do not touch the flat sides of the chromatography paper. Handle the paper by the edges, since oil from your fingers could distort the results.

❸ Use a pencil and ruler to draw a line 2 cm from the edge of one of the 22-cm sides. (See Figure 2.5.)

❹ Fold the 22-cm side of the paper in half, using a ruler to crease the paper sharply. Fold two more times and refold so as to make an accordion pleat as shown in Figure 2-5.

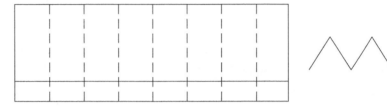

Figure 2-5: Chromatography paper with fold lines and accordion pleat

❺ Using a pencil, put a dot at the center of the 2-cm line on each of the four middle rectangles.

❻ Label the dots 1, 2, 3, and 4 with a pencil as shown in Figure 2-6.

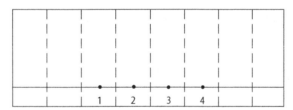

Figure 2-6: Chromatography paper with labeled dots

❼ Use separate capillary tubes to place small drops of each known compound to be tested onto the corresponding dot: dot 1 = ethyl vanillin, dot 2 = vanillin, dot 3 = coumarin, and dot 4 = 4-hydroxybenzaldehyde.

❽ Remove the plastic wrap from the beaker and stand the pleated paper in the beaker so that the 2-cm line is just above the alcohol solution. The alcohol solution must *not* be touching the line with the dots. Replace the plastic wrap and rubber band.

❾ Allow the beaker to stand undisturbed as the alcohol solution travels up the paper by capillary action. When the solution is about 2 cm from the top of the paper, remove the cover, take out the paper, and immediately mark the position of the solvent front using the pencil.

⑩ Wave the paper to dry off the alcohol solution.

⑪ Turn off the light(s) in the room and turn on the UV lamp. Do not look directly into the lamp. Place the paper under the UV lamp and circle the spots that fluoresce. Turn off the UV lamp.

⑫ Turn on the lights and mark the center of each circled spot.

⑬ Measure the distance from the 2-cm line to the top of the solvent front in each rectangle. Record this distance in Data Table 1 as D_f (distance of the solvent front).

⑭ Measure the distance from the 2-cm line to the center of each circled spot for the four samples. Record these distances as D_s (solute distance).

⑮ Calculate the R_f value for each of the known compounds using the formula in the following Calculations section.

⑯ Repeat the experiment using the commercial vanilla samples and record your data in Data Table 2. These samples may contain more than one of the known compounds.

Data Table 1 for "How Pure Is Your Vanilla?"			
	D_f	D_s	R_f
Sample #1 (ethyl vanillin)			
Sample #2 (vanillin)			
Sample #3 (coumarin)			
Sample #4 (4-hydroxybenzaldehyde)			

Data Table 2 for "How Pure Is Your Vanilla?"			
	D_f	D_s	R_f
Commercial sample #1			
Commercial sample #2			
Commercial sample #3			
Commercial sample #4			

Calculations

In paper chromatography, the paper is the stationary phase and the solvent is the mobile phase. As the mobile phase moves up the paper by capillary action, some components of the sample that was spotted onto the paper are strongly attracted to molecules in the mobile phase. They will then be carried along with the mobile phase. Other components that are less attracted to the mobile phase lag behind.

The distance traveled from the origin line by each component divided by the distance traveled by the mobile phase is called its R_f value.

$$R_f = \frac{D_s}{D_f}$$

Questions

❶ Compare the R_f values for each spot in the commercial vanilla samples with those of the known compounds. Determine which components are present in the vanilla samples.

_____

_____

_____

❷ Pure vanilla extract sold in the United States contains both vanillin and 4-hydroxybenzaldehyde. Can you identify which sample is a United States sample of pure vanilla extract?

_____

_____

_____

3 Some vanilla produced in Mexico contains coumarin. Coumarin has a flavor similar to vanilla beans. The FDA does not allow vanilla that contains coumarin to be sold in the United States. The Mexican vanilla typically sold at the airport is labeled "Pure vanilla extract, does not contain coumarin." The 500-mL bottle cost $1 U.S. Pure vanilla extract in the United States costs about $6 for a 60-mL bottle. What conclusions can you draw about these two types of vanilla extract?

_____

_____

_____

4 Look up the structural formulas for the four known samples. How are they similar and how are they different?

_____

_____

_____

Instructor Notes for Student Activity 2C

In Activity 2C, students use paper chromatography to separate the organic components of different commercial vanilla samples. By comparing the R_f values to those of known vanilla components, students can evaluate the purity of the commercial vanilla samples.

Safety and Disposal

It is your responsibility to review appropriate safety procedures with your students, including the use of personal protective equipment. No special disposal procedures are required.

Materials Notes

If rolls of chromatography ribbon are available instead of larger sheets of chromatography paper, the experiment can be modified by using separate test tubes and stoppers or beakers covered with parafilm. In this case a separate strip of chromatography paper is used for each sample. The standard compounds used for comparison may be obtained from Sigma Chemical Co., P.O. Box 14508, St. Louis, MO 63178; 800/325-3010, *http://www.sigma-aldrich.com.*

Answers to Questions

❶ *Compare the R_f values for each spot in the commercial vanilla samples with those of the known compounds. Determine which components are present in the vanilla samples.*

Values will vary depending on the vanilla samples used. Coumarin can no longer be included in vanilla sold in the United States because it has been associated with liver cancer. However, it is occasionally found in Mexican vanilla that was not intended for importation.

❷ *Pure vanilla extract sold in the United States contains both vanillin and 4-hydroxybenzaldehyde. Can you identify which sample is a United States sample of pure vanilla extract?*

Answers will vary depending on vanilla samples used.

❸ *Some vanilla produced in Mexico contains coumarin. Coumarin has a flavor similar to vanilla beans. The FDA does not allow vanilla that contains coumarin to be sold in the United States. The Mexican vanilla typically sold at the airport is labeled "Pure vanilla extract, does not contain coumarin." The 500-mL bottle cost $1 U.S. Pure vanilla extract in the United States costs about $6 for a 60-mL bottle. What conclusions can you draw about these two types of vanilla extract?*

From their experimental data, students may be able to conclude that Mexican vanilla may contain other compounds that allow it to be less expensive.

❹ *Look up the structural formulas for the four known samples. How are they similar and how are they different?*

Coumarin

Vanillin
(4-hydroxy-3-methoxybenzaldehyde)

Ethyl vanillin
(3-ethoxy-4-hydroxybenzaldehyde)

4-hydroxybenzaldehyde

Each substance has a benzene ring, an alcohol and/or ether functional group, and an aldehyde or ketone functional group. The alcohol and aldehyde groups on three of the molecules (vanillin, ethyl vanillin, and 4-hydroxybenzaldehyde) are in the para position.

Coumarin is a benzopyranone. The other three substances are derivatives of benzaldehyde and have a hydroxy group (–OH) on the number 4 carbon. Two of these have a methoxy or ethoxy substitution on carbon number 3.

Section 3

The Production Process—From Blending to Freezing

A Step-by-Step Overview of Ice Cream Production (Student Background 3)

Commercial ice cream plants vary in size—from small, family-run operations that produce individual batches largely by hand, to giant, highly mechanized and automated plants characterized by continuous operation. However, all ice cream production processes have certain basic steps in common: composing the ice cream mix by blending the ingredients, homogenizing and pasteurizing, cooling, aging (in a storage tank), freezing, packaging, hardening, storing, and finally loading out the finished product. Laboratory technicians assure the quality of the ice cream by carefully monitoring each step of the process. The flow chart in Figure 3-1 illustrates the steps in manufacturing ice cream.

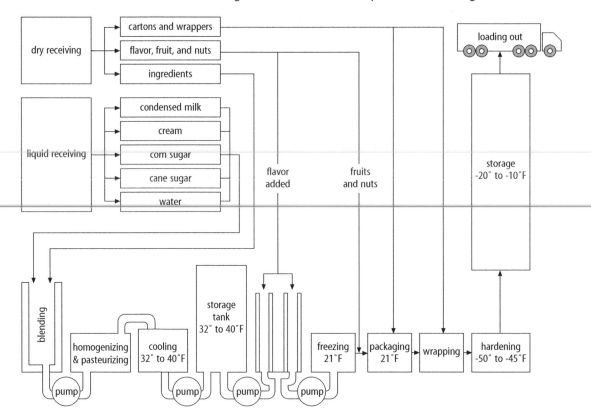

Figure 3-1: Ice cream manufacturing process

Composing the mix from raw ingredients is the first step in ice cream manufacturing. This step involves moving the ingredients from the receiving or storage area, measuring out the desired amounts, and blending them together.

Liquid ingredients, such as milk, cream, and syrups, are stirred together and heated in a vat. Dry ingredients, such as dried eggs, cocoa, sugar, and stabilizers, are added slowly as the liquids are stirred together.

Federal law requires the ice cream mix to be pasteurized to destroy disease-producing bacteria. This process also aids in the dissolving and blending of the mix, improves the flavor, and improves the uniformity of the product. Pasteurization involves rapidly raising the temperature of the mix to a predetermined point, holding it constant at that temperature for a given time, then cooling it rapidly to below 40°F (4°C). The amount of heating time varies with the temperature, as shown in the following table.

Grade "A" Pasteurized Milk Ordinance (1999) Recommended Minimum Times and Temperatures for Pasteurization of Ice Cream Mixes		
Method	Time	Temperature
Batch	30 minutes	150°F (66°C)
HTST (high temperature, short time)	15 seconds	166°F (75°C)
UHT (ultra-high temperature)	Determined by a process authority	Determined by a process authority

Homogenizing the ice cream mix creates a permanent and uniform suspension of fat globules in the mix by reducing their size to less than 2 μm. If the mix is not properly homogenized, the fat rises and forms a cream layer, giving the frozen product a greasy appearance and feel. Homogenization is accomplished by forcing the ice cream mix through a very small opening under high pressure.

After the mix has been pasteurized and homogenized, it is aged in a storage tank for at least 4 hours. Aging improves the smoothness and texture of the final product.

Liquid and pureed flavorings are generally added to the mix just prior to freezing; solid materials such as nuts and candies are mixed into the soft frozen product as it comes out of the freezer. Although some manufacturers have as many as 500 different flavoring recipes, vanilla and chocolate still dominate the market. Before the ice cream mix enters the freezer, it is tested to determine whether the mix meets formula specifications.

To begin the freezing process, the mix is pumped into a freezer barrel chilled with a liquid refrigerant as shown in Figure 3-2.

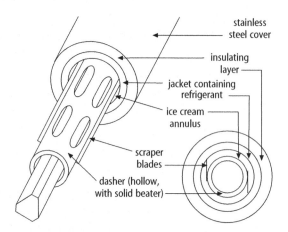

Figure 3-2: Continuous barrel freezer

As the mixture cools rapidly, dashers whip air into the mixture and scrape tiny ice crystals from the cylinder walls. Gradually the liquid mix turns into a viscous foam containing suspended fat globules, colloidal proteins, carbohydrates, and salts intertwined with ice crystals and small pockets of trapped air. In the trade, these pockets are often referred to as air cells. Figure 3-3 contains a diagram showing a microscopic representation of this mixture.

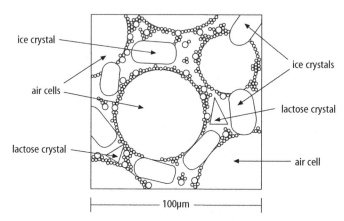

Figure 3-3: Microscopic representation of ice cream structure

The more quickly the ice cream freezes, the smaller the ice crystals that form, and the smoother the texture of the ice cream. Slower freezing allows the ice crystals to increase in size, which causes the ice cream to have a coarse, grainy texture. Freezing times depend on physical and mechanical factors such as the type and condition of the freezer, the speed of the dasher, the temperature and velocity of the refrigerant, and the composition of the mix.

Continuous freezers are self-contained units that may be manually operated but often are completely computerized. They are available with one, two, or three barrels and can produce between 100 and 1,200 gallons of ice cream per hour.

As ice cream leaves the freezer, it is packaged in one of two ways: in bulk packaging for the sale of dipped products such as ice cream cones, or in consumer packaging for direct retail sale. Filling machines are used to automatically package 2½- to 5-gallon bulk containers and consumer packages as small as 3 fluid ounces (single-serve cups) up to 5-gallon pails. The most common consumer package contains ½ gallon. Filling machines must give a precise and accurate fill, avoid air pockets, maintain even distribution of inclusions such as fruits and nuts and, above all, meet strict sanitary standards.

In the filled containers, ice cream is still only semisolid and must be hardened. Packages are stored in a hardening facility until the temperature at the center of the package reaches −0.4°F (−18°C) or lower. Again, the more quickly this happens, the smoother the product will be. Hardening time of 6–8 hours for a 5-gallon package produces an excellent product. The product is then stored prior to loading out to consumers.

Transferring the final product from the plant to the frozen food section in your grocery store requires refrigerated trucks and careful handling so that the ice cream reaches the consumer at its peak of quality. Any thawing and refreezing during loading out and subsequent storage causes ice crystal growth, which reduces the quality of the ice cream.

● Plant Tour (Overhead 3A)

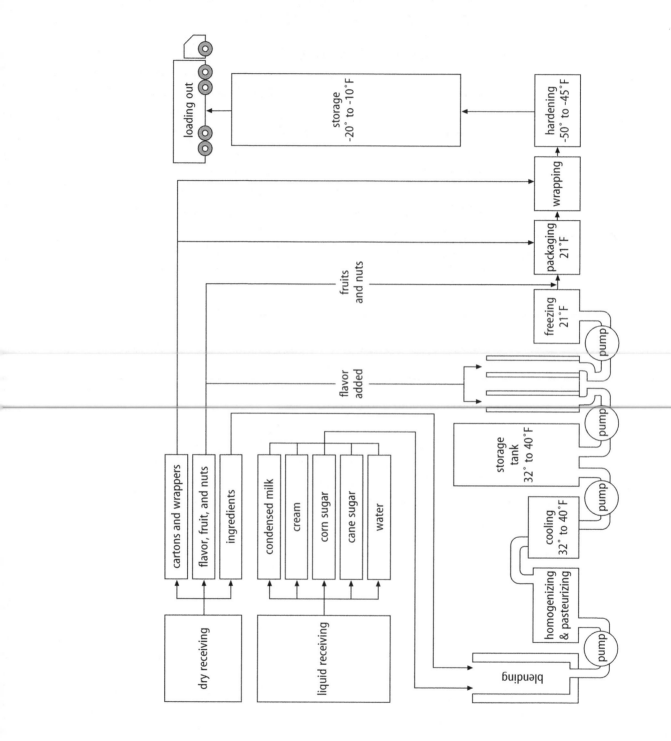

● Continuous Freezer (Overhead 3B)

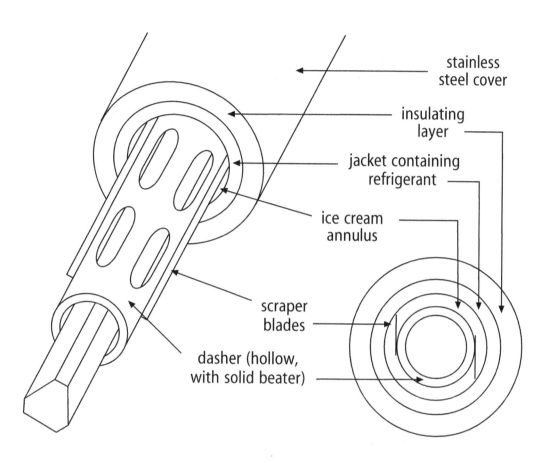

stainless steel cover

insulating layer

jacket containing refrigerant

ice cream annulus

scraper blades

dasher (hollow, with solid beater)

● The Effect of Sweeteners on Freezing Point (Student Activity 3)

Removal of sufficient heat from an ice cream mix to drop the temperature of the mix below its freezing point causes ice crystals to form. The freezing point of an ice cream mix depends on the concentration of the dissolved constituents, especially the sweetener. Because sugars do not ionize in solution, their effect on the freezing point varies inversely with their molecular weight, given the same concentration. In this experiment you will investigate the effect of two common ice cream sweeteners, sucrose and dextrose, on the freezing point of water.

Materials
Per pair of students
- wax pencil or permanent marker
- 100-mL graduated cylinder
- 6 Styrofoam® cups
- 6 paper cups (3-ounce)
- thermometer
- 60 g dextrose
- 60 g sucrose
- 2 small trays or boxes

Per class
- centigram balances

Safety and Disposal
As instructed by your teacher, follow appropriate safety procedures, including the use of personal protective equipment such as goggles and an apron. No special disposal procedures are required.

Procedure
Day 1

❶ Measure 100 mL distilled water using the graduated cylinder. Pour 100 mL water into a Styrofoam cup. Mark the 100 mL level on the side of the Styrofoam cup. Repeat the procedure with the other five Styrofoam cups. Label the Styrofoam cups 10% dextrose, 20% dextrose, 30% dextrose, 10% sucrose, 20% sucrose, and 30% sucrose.

❷ Measure 50 mL distilled water using the graduated cylinder. Pour 50 mL water into a paper cup. Mark the 50 mL level on the side of the paper cup. Repeat the procedure with the other five paper cups. Label the paper cups 10% dextrose, 20% dextrose, 30% dextrose, 10% sucrose, 20% sucrose, and 30% sucrose.

❸ Prepare a 10% dextrose solution by placing 10 g dextrose in the appropriately labeled Styrofoam cup and adding enough distilled water to reach the 100 mL line.

❹ Prepare a 20% dextrose solution by placing 20 g dextrose in the appropriately labeled Styrofoam cup and adding enough distilled water to reach the 100 mL line.

❺ Prepare a 30% dextrose solution by placing 30 g dextrose in the appropriately labeled Styrofoam cup and adding enough distilled water to reach the 100 mL line.

❻ Prepare 10, 20, and 30% sucrose solutions in the same manner.

❼ Pour 50 mL of each of the six solutions into appropriately labeled paper cups by filling the cups up to the 50 mL line. Place the paper cups on a small tray or in a small box in the freezer portion of the refrigerator. Place the Styrofoam cups with the other half of the solutions on a small tray or in a small box in the fresh food portion of the refrigerator.

Day 2

❶ Remove one of the cups from the refrigerator and its corresponding cup from the freezer.

❷ Tear open the paper cup and place the frozen portion in the Styrofoam cup containing its corresponding liquid.

❸ Using the thermometer to gently stir the mixture, record the equilibrium temperature of the liquid-solid mixture. (The equilibrium temperature should be recorded when both solid and liquid are present and no further drop in temperature is observed).

❹ Repeat steps 1–3 with the remaining cups.

❺ Plot the freezing points of the sucrose and dextrose solutions as a function of their concentration.

Questions

❶ How does the concentration of the sweetener affect the freezing point of the solution?

_____

_____

_____

❷ Which sweetener has a greater effect on the freezing point of the solution? (You might want to consider the following questions to answer this.)

_____

_____

_____

a. Were the solutions of dextrose and sucrose prepared differently?

_____

b. What could account for the differences observed?

_____

c. Look up the chemical formulas for these two sugars. Are they the same?

_____

d. Calculate the molecular weights of the two sugars. Which molecular weight is less?

_____

e. Look back to how you prepared the 10% solutions. What did you do?

_____

f. How many moles of dextrose were in the 10% solution? (The molecular weight of dextrose is 180 g/mole.)

_____

g. How many moles of sucrose were in the 10% solution? (The molecular weight of sucrose is 342 g/mole.)

_____

h. How does the number of moles of the different sugars affect the observed freezing point?

_____

❸ Look in your chemistry textbook for the definition of colligative properties. How does this activity demonstrate that the freezing point depression is a colligative property?

_____

_____

_____

Although any solute will tend to depress the freezing point of water by interfering with the formation of ice crystals, it is the number of solute particles present that controls the amount of depression observed. When ice cream freezes, the water portion of the mix forms ice crystals that are practically pure water. As temperature continues to decrease, the sugar, which constitutes the major solute, becomes increasingly more concentrated in the remaining liquid water and thus continues to lower the freezing point of the mixture even further.

Sucrose, the most commonly used sweetener in ice cream manufacture, is highly soluble; its effect on the freezing point of water is significant. An increase of 2% in sucrose content lowers the freezing point approximately 0.23°F (0.13°C). Other sweeteners are often compared to sucrose in their ability to depress the freezing point, with sucrose being given a freezing point equivalence factor of 1.00. Dextrose, or refined corn sugar, is a monosaccharide and lowers the freezing point more than sucrose, with a relative freezing point depression factor of 1.90. Accordingly, 2% dextrose lowers the freezing point of water by approximately 1.90 multiplied by 0.23°F, which equals 0.44°F.

Safety and Disposal

It is your responsibility to review appropriate safety procedures with your students, including the use of personal protective equipment. No special disposal procedures are required.

Procedure Notes

If a refrigerator with a freezer is not available in the science area, arrangements may be made to use freezers in the cafeteria or the family/consumer science department. This experiment requires a portion of each of two days, as part of the sample must be frozen overnight. Alternatively, you could prepare the samples in advance and have the students carry out only Day 2 of the experiment.

Sample Results Table

Sample Results for "The Effect of Sweeteners on Freezing Point"			
Sucrose	Freezing Point	Dextrose	Freezing Point
10%	30.6°F (−0.8°C)	10%	29.3°F (−1.5°C)
20%	29.3°F (−1.5°C)	20%	27.0°F (−2.8°C)
30%	27.5°F (−2.5°C)	30%	23.4°F (−4.8°C)

Answers to Questions

❶ *How does the concentration of the sweetener affect the freezing point of the solution?*

As the concentration of the sweetener increases, the freezing point becomes lower.

❷ *Which sweetener has a greater effect on the freezing point of the solution? (You might want to consider the following questions to answer this.)*

a. *Were the solutions of dextrose and sucrose prepared differently?*

The amounts used and preparation procedures were the same, but the sugars used were different.

b. *What could account for the differences observed?*

The two chemicals are different sugars.

c. *Look up the chemical formulas for these two sugars. Are they the same?*

No. Dextrose is $C_6H_{12}O_6$. Sucrose is $C_{12}H_{22}O_{11}$.

d. *Calculate the molecular weights of the two sugars. Which molecular weight is less?*

The molecular weight of dextrose is 180 g/mole and the molecular weight of sucrose is 342 g/mole. Dextrose has the lesser molecular weight.

e. *Look back to how you prepared the 10% solutions. What did you do?*

The solutions were prepared by mixing 10 g of each sugar with water to make a 100 mL solution.

f. *How many moles of dextrose were in the 10% solution? (The molecular weight of dextrose is 180 g/mole.)*

0.055 moles dextrose

g. *How many moles of sucrose were in the 10% solution? (The molecular weight of sucrose is 342 g/mole.)*

0.029 moles sucrose

h. *How does the number of moles of the different sugars affect the observed freezing point?*

The greater number of moles of sugar added, regardless of type, lowers the resulting freezing point and increases the freezing point depression observed.

Returning to the original question:

Dextrose has a greater effect on the freezing point than sucrose does due to its lower molar mass.

❸ *Look in your chemistry textbook for the definition of colligative properties. How does this activity demonstrate that the freezing point depression is a colligative property?*

For colligative properties, the effect is relative to the molal concentration of the solute, so with given weights of solute and volume of solvent, the effect on the freezing point is greater for sugars with lower molecular weights. The other colligative properties of solutions are boiling point (raised by solutes), vapor pressure (lowered by solutes), and osmotic pressure (raised by solutes).

Section 4

Quality Assurance—The Role of the Laboratory Technician in Product Testing

● U.S. FDA Standards for Ice Cream Manufacturing (Student Background 4A)

Government classification and regulation of the ice cream industry began in 1910–1911, when a classification was adopted to indicate whether or not eggs were present in an ice cream product. Over the years, various classification systems have been used to describe ice cream. The most recent U.S. Food and Drug Administration (FDA) 21 CFR 135 standards regulating the content of ice cream took effect in September 1995. Government standards define common ice cream products in the following manner.

Ice cream: A product containing at least 10% milkfat and 20% total milk solids, safe and suitable sweeteners, and optional stabilizing and flavoring ingredients. Finished ice cream must weigh at least 4½ pounds per gallon and contain a minimum of 1⅗ pounds of food solids per gallon.

Frozen custard, French ice cream: Ice cream that contains a minimum of 1.12–1.40% egg yolk solids by weight.

Reduced-fat ice cream: Ice cream made with 25% less fat than the reference ice cream.

Light (or "lite") ice cream: Ice cream made with 50% less fat or containing one-third fewer calories than the reference ice cream.

Low-fat ice cream: Ice cream that has 3.0 g or less milkfat per serving. (For ice cream, 4 fluid ounces weighing at least 64 g is considered one serving.)

Nonfat ice cream: Ice cream containing less than 0.5 g milkfat per serving.

Soft-serve ice cream: Ice cream served directly from the freezer without hardening.

Frozen yogurt: Similar to ice cream but lower in fat content. It must contain yogurt bacteria of the species *Lactobacillus bulgarius* and *Streptococcus thermophilus*. Titratable acidity is generally 0.3–0.5%, depending on the amount of NMSs it contains.

Fruit sherbet: Contains fruit juices, sugar, stabilizers, 1.2% milkfat, and 1.3% NMSs. Acidity must be at least 0.35%.

Water ice or ice: A mixture of fruit juice, sugar, and stabilizers but no dairy products. May be frozen with or without agitation.

The Importance of Air in Ice Cream (Student Background 4B)

As the dasher whips the ice cream in the freezer, air becomes incorporated into the mix and the overall volume increases. The increase in volume of ice cream over the volume of the mix used is known as overrun. The amount of overrun influences quality and profits and is subject to legal standards.

The composition of the ice cream mix and the way it is processed determines the amount of air that should be incorporated into the mix. Too much air produces fluffy, snowlike ice cream; too little air leads to a heavy, almost soggy product. Some of the factors to consider when determining the appropriate percentage of overrun include government regulations; the total solids content of a mix (a mix with higher total solids content may incorporate more air); type of ice cream (fruit and nut additions require lower overrun); packaging (bulk packages sold for dipping contain higher overrun); and finally, the projected selling price of the ice cream.

During the freezing process, the sharpness of the scraper blades, speed and design of the dasher, and temperature and volume of refrigerant all play important roles in obtaining the desired amount of overrun. Some of the mix properties that depress overrun include high fat or corn syrup solids content, excessive amounts of stabilizers and calcium salts, and the presence of fruits or cocoa. Mechanical factors that depress overrun include dull freezer blades and freezing the mix too stiff. Factors that enhance overrun include the addition of sodium caseinate, whey solids, egg yolks, and certain emulsifiers and stabilizers. Pasteurizing the mix at a higher temperature also increases overrun. Properly operating continuous freezers, with their pressurized cylinders and very sharp blades, produce adequate overrun with nearly all mix formulations. The batch freezer, however, which operates under atmospheric pressure, depends heavily on the whipping ability of the mix to determine how much air is incorporated into the frozen mix.

Keeping the amount of overrun as consistent as possible from batch to batch and from day to day produces a uniform product that will gain customer approval. Variation of 10% in overrun represents a remarkable difference in profits and can certainly be observed as a difference in quality by the consumer. Different ice cream products have different general amounts of overrun as shown in the following table.

Percentage Overrun for Ice Cream Products	
Product	% Overrun
Ice Cream, Packaged	70–100%
Ice Cream, Bulk	90–100%
Sherbet	30–50%
Soft-Serve Ice Cream	50–70%
Superpremium Ice Cream	20–40%

Overrun is usually highest when artificial flavoring is added instead of natural flavoring. Artificial flavorings are generally used in the economy types of ice cream, and economy is achieved by selling as little ice cream mix as is permitted by law. Natural flavorings and relatively low overrun characterize higher-priced ice creams such as Edy's® or Breyers®.

The ice cream industry places flavorings into three categories as follows: Category I—only pure natural flavoring is used; Category II—pure flavoring predominates over artificial; Category III—artificial flavoring predominates or is the only flavoring.

● How "Light" Is Your Ice Cream? (Student Activity 4A)

Before ice cream leaves the plant, the laboratory technician performs a series of tests to monitor the quality of the final product. One of these tests measures the percent overrun. Overrun is defined as the increase in the volume of finished ice cream divided by the volume of mix used. It is caused by the addition of air beaten into the ice cream mix as it freezes and by the expansion of water as it freezes. In this activity you will calculate the percent overrun in several varieties of ice cream.

While "light" or "lite" often refers to ice cream with reduced fat or calories, lightness in terms of an airy product is a measure of overrun. A fluffy texture with large air cells present results from incorporating a large amount of air during the freezing process. The incorporated air changes the density of the final product. Typically superpremium ice creams have less overrun and are the most dense, while generic or store brands have the most overrun and have a more fluffy or foamy texture. The amount of overrun can also be observed qualitatively by the amount of foam that forms when the ice cream melts.

Materials

Per pair of students
- ice cream sample
- 250-mL beaker
- 600-mL beaker
- hot plate
- thermometer

Per class
- centigram balances

Safety and Disposal

As instructed by your teacher, follow appropriate safety procedures, including the use of personal protective equipment such as goggles and an apron. Ice cream used in a laboratory setting must NEVER be tasted.

No special disposal procedures are required.

Procedure

❶ Weigh the 250-mL beaker and record its weight in the data table.

❷ Prepare a warm water bath by filling the 600-mL beaker approximately half full with water and heating it on the hot plate to about 50°C.

❸ Obtain an ice cream sample from your teacher in the preweighed 250-mL beaker. Weigh and record the weight of the beaker and sample.

❹ Calculate the weight of the ice cream sample.

❺ Place the beaker containing the ice cream sample into the warm water bath, keeping the water bath on the hot plate at a low setting for 15–20 minutes or until the ice cream melts and a distinct layer forms between the liquid mix at the bottom of the beaker and the foam on top.

❻ Group all of the class samples in a row, separated by brand, and observe and record the approximate amount of foam present in each brand compared to the liquid mix present in the sample.

❼ Assuming that your ice cream sample had a volume of 8 fluid ounces and that a standard ice cream mix weighs 31.9 g/fluid ounce, calculate the mass of the mix that would be present if the sample had no air beaten into it.

❽ Using the actual mass of your frozen ice cream sample, calculate the overrun in your sample:

$$\% \, overrun = \frac{(grams \, of \, mix - grams \, of \, frozen \, ice \, cream)}{grams \, of \, frozen \, ice \, cream} \times 100$$

❾ Compile the class data for each brand of ice cream tested and calculate the average overrun for each type of ice cream.

Data Table 1 for "How 'Light' Is Your Ice Cream?"	
brand of ice cream sample tested	
mass of 250-mL beaker (empty)	
mass of 250-mL beaker and ice cream sample	
mass of ice cream sample	
% overrun of your sample	

Data Table 2 for "How 'Light' Is Your Ice Cream?"		
	Class Average % Overrun for Various Brands Tested	Cost of Various Brands Tested
store brand or generic brand		
trade brand		
superpremium brand		

Questions

1 How does the amount of overrun compare for the various brands of ice cream tested?

_____

_____

_____

2 How did the amount of foam present in the samples compare to the amount of overrun for the various brands of ice cream tested?

_____

_____

_____

3 How does the cost compare to the amount of overrun for the various brands of ice cream tested?

_____

_____

_____

Instructor Notes for Activity 4A

In Activity 4A the students compare the mass of 8 fluid ounces of several types of ice cream, including an inexpensive store brand, a trade brand, and a superpremium brand, calculating the overrun present in each case.

Safety and Disposal

It is your responsibility to review appropriate safety procedures with your students, including the use of personal protective equipment. Ice cream used in a laboratory setting must NEVER be tasted.

No special disposal procedures are required.

Materials Notes

A half gallon of ice cream will provide samples for eight lab groups. Depending on the class size, each group of students will examine one or more samples so that eight samples of both the store brand and the trade brand have been tested. Because of cost, it is practical to have only two groups investigate the superpremium brand—this is generally sold in pint containers so splitting the pint in half will allow each group an 8-fluid-ounce sample. Use vanilla ice cream for all samples. Labeling will be "artificially flavored vanilla" for the store brand, "vanilla flavored" for the trade brand, and "vanilla" for the superpremium brand.

Procedure Notes

Keep the ice cream in a deep freeze or in a cooler packed with dry ice. When the students are ready to obtain their samples, open the rectangular half-gallon container, completely exposing all but one side of the ice cream. Cut the product precisely in the center with a warm knife. (Dip the knife blade into hot water.) Cut the halves and then the fourths again, forming eight pieces of equal size. For the superpremium, split the pint into two equal portions. Students can then assume that their sample has a volume of 8 fluid ounces. A standard ice cream mix has a mass of 4,080 g/gallon or 31.9 g/fluid ounce. Although the mass of various mixes may vary slightly, for the purposes of this activity a standard mass of 31.9 g/fluid ounce may be assumed. If there were no overrun, the mass of the mix to fill 8 fluid ounces would be 255 g. The difference between this and the actual mass of the frozen ice cream represents the amount of air beaten into the ice cream during the freezing process.

The activity could be extended to investigate the difference in overrun for various flavors or for sherberts or ices.

Sample Results

The amount of overrun will vary, depending on the brand and type of the ice cream. Generally, superpremium brands will show around 20% overrun, while trade brands will be around 80%, and inexpensive store brands may be close to 100% overrun.

Answers to Questions

❶ *How does the amount of overrun compare for the various brands of ice cream tested?*

The superpremium ice cream has the least overrun; the store brand has the most; and the trade brand falls in between but generally closer to the store brand than to the superpremium brand.

❷ *How does the amount of foam present in the samples compare to the amount of overrun for the various brands of ice cream tested?*

When the superpremium ice cream melts, only a small layer of foam appears at the top of the melted mix. With the trade brand and the store brand, there is a layer of foam on top of the melted mix that is almost equal in volume to the amount of mix itself.

❸ *How does the cost compare to the amount of overrun for the various brands of ice cream tested?*

It appears that overrun is inversely related to price, with the superpremium brand having by far the highest price and the lowest amount of overrun.

● Quality Equals Safety (Student Background 4C)

The laboratory technician performs standard plate counts (SPCs) on the raw dairy products and standard plate and coliform counts on the finished product to monitor the microbiological quality of ice cream. The generally accepted maximum coliform count on the finished product is 10 coliform bacteria/g. However, many food science industry personnel maintain that no detectable coliform bacteria should be present in a 1-g sample of ice cream. These bacteria are readily killed by pasteurization and any presence of a viable coliform bacterium indicates that the mix was recontaminated after pasteurization, making possible the presence of disease-producing bacteria.

SPCs are designed to determine the total aerobic bacterial population. Given that fresh raw milk samples today generally contain fewer than 50,000 total aerobic bacteria/mL and that only a few of those are able to survive pasteurization, a generally accepted standard for an SPC is less than 500 colonies/g. The bacteria that survive pasteurization are nearly all spore-forming types.

Because pasteurization of the mix is the critical step in bacterial control, this step must be carefully monitored. Ingredients added following pasteurization are a potential introduction point for bacteria. Fruit pieces, nut meats, or any natural additive such as cinnamon or vanilla beans may contain microorganisms that will lead to elevated SPCs and coliform counts.

To ensure that ice cream meets such high microbial quality standards, employees in the ice cream industry must be vigilant that all environmental contaminants are removed from all food contact surfaces that touch the ingredients and products prior to packaging. Sanitary ice cream production requires clean, healthy, and sanitation-minded employees; high-quality raw ingredients; proper processing; immaculately clean and sanitary equipment; and finally, proper storage and distribution of the product.

● How Many Microbes Are Present? (Student Activity 4B)

Bacteria present in ice cream may originate from the original ingredients and from handling before or after pasteurization. Laboratory technicians regularly take bacteria counts on both incoming milk and finished ice cream. Since the pasteurization step is designed to kill disease producers and reduce bacterial counts, safe handling of the mix following pasteurization is particularly important. Thorough cleaning and sanitizing of all equipment cannot be overemphasized if a plant is to produce ice cream free of objectionable bacteria. In this experiment you will carry out a standard plate count (SPC) designed to determine the total aerobic bacterial population present in several commercial ice cream samples.

Materials

Per pair of students (for each kind of ice cream tested)
- 20 g ice cream
- sterile spatula
- large, sterile test tube (25 mm x 200 mm)
- 10-mL sterile pipet
- sterile 99-mL dilution bottle
- sterile 9-mL dilution bottle
- 3 sterile 1-mL pipets
- 3 Petrifilm™ Aerobic Count Plates

Per class
- standard colony counter or other illuminated magnifier
- several water baths
- centigram balances

Safety and Disposal

As instructed by your teacher, follow appropriate safety procedures, including the use of personal protective equipment such as goggles and an apron. After use, the Petrifilm will contain viable bacteria colonies. Do not lift the film or handle the plates unnecessarily. Return the plates to your teacher for disposal. Always wash hands thoroughly after working with microbiological materials.

Procedure (for each ice cream sample to be tested)
Part A: Preparation of the sample

❶ Using a sterile spatula, transfer about 20 g of the ice cream sample to the large sterile test tube and place the test tube in a 113°F (45°C) water bath until the ice cream is melted (not more than 15 minutes).

❷ Pour 99 mL sterile distilled water into the 99-mL dilution bottle. Place the bottle on the balance.

❸ Mix the melted ice cream with the 10-mL pipet and transfer 11 g ice cream directly into the dilution bottle. Cap and shake well. This is a 1:10 dilution, which will be referred to as the 10^{-1} dilution.

❹ Pour 9 mL sterile distilled water into the 9-mL dilution bottle.

❺ Using a 1-mL pipet, transfer 1 mL of the 1:10 dilution into the 9-mL dilution bottle. Cap and shake well. This is a 1:100 dilution, which will be referred to as the 10^{-2} dilution.

Part B: Inoculation of Petrifilm

❶ Place the Petrifilm Aerobic Count Plate on a flat surface. Peel open the Petrifilm plate, being careful not to touch the nutrient gel with your fingers. (See Figure 4-1a.)

❷ Fill the 1-mL pipet with the 10^{-1} diluted sample. With the pipet perpendicular to the plate, place 1 mL of the 10^{-1} diluted sample onto the center of the bottom film. (See Figure 4-1b.)

❸ Release the top film, allowing it to *drop*—do not roll the top film down. (See Figure 4-1c.)

❹ With the ridge side down, place the spreader on the top film over the inoculum. (See Figure 4-1d.)

❺ Apply *gentle* pressure on the spreader to distribute inoculum over the circular area. Do not twist or slide the spreader. (See Figure 4-1e.)

❻ Lift the spreader. Wait at least 1 minute for the gel to solidify. (See Figure 4-1f.)

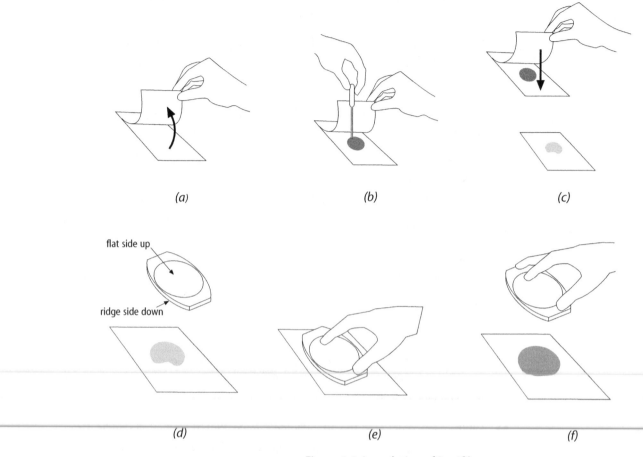

(a)   *(b)*   *(c)*

flat side up

ridge side down

(d)   *(e)*   *(f)*

Figure 4-1: Inoculation of Petrifilm

❼ Repeat steps 1–6 with the 10^{-2} dilution and a separate Petrifilm plate.

❽ Incubate the plates 48 hours ± 3 hours at 95°F (35°C).

❾ Count the colonies as instructed and record your results, along with the brand and flavor of the sample.

Data Table for Inoculation of Petrifilm	
	Total Colonies
Sample 1:	
10^{-1} dilution	
10^{-2} dilution	
Sample 2:	
10^{-1} dilution	
10^{-2} dilution	
Sample 3:	
10^{-1} dilution	
10^{-2} dilution	

Questions

❶ Does the SPC for the ice cream samples you examined fall within generally accepted guidelines?

_____

_____

_____

❷ What differences in SPC do you observe for the different ice creams tested, and what explanations can you offer for these observed differences?

_____

_____

_____

❸ What factors might lead to a high SPC, and how might a manufacturer control them?

_____

_____

_____

Instructor Notes for Activity 4B

In Activity 4B, pairs of students measure the microbiological counts of three different samples by performing a standard plate count (SPC) to determine the total aerobic bacterial population.

Safety and Disposal

It is your responsibility to review appropriate safety procedures with your students, including the use of personal protective equipment. Students must observe aseptic procedures to avoid introducing contaminating bacteria during the Procedure. For advice on how to set up a sterile experiment, a physical science teacher may want to consult with a life science teacher. To check their sterile technique, have students run a set of control plates. Prepare the control plates by inoculating several plates after having made the same dilutions with sterile distilled water in place of ice cream.

After use, the plates will contain viable bacterial colonies. Do not separate or handle the plates unnecessarily. Soak the plates in 10% bleach solution or autoclave them before disposing of them in the garbage. Always wash hands thoroughly after working with microbiological materials.

Materials Notes

This laboratory activity is written for the use of 3M™ Petrifilm™ Aerobic Count Plates which can be obtained through Flinn Scientific, P.O. Box 219, Batavia, IL 60510; 800/452-1261. Details for use and instructions for counting colonies are provided with the Petrifilm, and 3M will supply technical assistance to educators using their products (800/328-6553). Alternatively, the laboratory may be carried out using Plate Count agar (Standard Methods agar) poured into petri dishes.

Sample Results

Data will vary, depending on the ice cream used, but the SPCs your students obtain in the activity should be below 500 colonies/g. Samples containing fruit or nut pieces often will have elevated plate counts due to the presence of bacteria in those materials, which are added to the mix following pasteurization.

Answers to Questions

1 *Does the SPC for the ice cream samples you examined fall within generally accepted guidelines?*

Answers will vary depending on the samples examined.

2 *What differences in SPC do you observe for the different ice creams tested, and what explanations can you offer for these observed differences?*

As previously noted, the presence of fruit or nut additives may explain higher plate counts. Cinnamon, vanilla pieces, and some natural colorings also may increase the plate count.

3 *What factors might lead to a high SPC, and how might a manufacturer control them?*

Other factors leading to elevated SPCs include unsanitary equipment, careless handling of materials during processing, unsanitary surrounding such as poor ventilation, and poor hygienic practices on the part of employees.

Manufacturers control for microbial contamination by thoroughly washing and sanitizing the equipment. The pasteurization process is used to eliminate microbial contamination in the ice cream mix itself. Because fruits, nut meats, or other items are added after the pasteurization process, microbe counts might be higher in ice cream containing these additives.

Section 5

Ice Cream Take-Out

● Melting Behavior of Ice Cream (Student Research Investigation 5A)

Ideally, melted ice cream should resemble the original, unfrozen mix. The actual appearance of the melted ice cream, however, may be distinctly different than that of the original mix. One of the most common differences is the appearance of a great deal of foam in the melted product. High overrun due to large amounts of air incorporated into the ice cream will produce a foamy product. An excessive amount of egg solids will also contribute to a foamy appearance. Breakdown (denaturation) of proteins in the ice cream, caused by factors such as excess acidity, melting and refreezing, and long storage at low temperatures, will cause the meltdown to appear "curdy."

When ice cream does not appear to melt as it warms or seems to melt unusually slowly, it will also appear gummy or sticky. Stabilizers such as various gums and carrageenan (obtained from seaweed) form gel structures with some of the water in the ice cream to cause this gummy appearance. Stabilizers prevent the formation of large ice crystals when the ice cream melts and refreezes; however, slow melting indicates overstabilization of the mix.

Fat content also affects melting behavior so that low-fat ice cream will have distinctly different melting characteristics than superpremium samples.

Research Assignment: Examine the "melting behavior" of various commercial ice cream samples. Assess the characteristics of the melted sample and the rate at which the sample melts.

Instructor Notes for Student Research Investigation 5A

Consumer purchasing decisions and satisfaction with ice cream are based on price and quality. In recent years, the popularity of superpremium ice creams demonstrates consumers' willingness to pay a higher price for higher quality.

In the Research Investigations, based on the background information provided, students design experiments to assess the characteristics of commercial ice cream samples of varying quality and price. In industry, laboratory technicians assess these characteristics to determine the quality of the final product before it leaves the plant. Each of the Research Investigations is designed to be conducted outside of the classroom.

In Investigation 5A, students should test a variety of commercial ice cream products for their melting behavior. A standard meltdown test places a specific amount, such as 4 fluid ounces, of ice cream on wire gauze at 60°F (16°C). The time to collect 10 mL of liquid is recorded, then the amount of liquid collected is measured each 10 minutes. A graph of milliliters of meltdown per hour can then be plotted. Students may wish to photograph the procedure at timed intervals to assess retention of shape and the nature of the melted mix. Ice creams of different price ranges should be examined along with those of different fat contents. Frozen yogurt will provide an interesting comparison due to the high acidity of the yogurt mix.

● Amount and Distribution of Mix-Ins (Student Research Investigation 5B)

Although vanilla and chocolate "lead the pack" with about 50% of the market share, many other flavors are also available. Some flavors include mix-ins, which are particles that provide flavor and alter texture. These mix-ins vary from common fruits and nuts to exotic pieces of candy and morsels of bubble gum. When preparing an ice cream with larger-sized particles mixed in, the amount used and uniform distribution throughout the batch are important considerations.

In some instances, the name of the ice cream flavor is not totally representative of the mix-in, as in the case of chocolate chip. Many inexpensive brands with that name add chocolate syrup, which freezes into tiny flakes as it is added to the semifrozen ice cream, rather than adding traditional chocolate chips.

The amount of mix-in clearly will affect the flavor of the ice cream and is a factor that the manufacturer must consider when consumer acceptance of the product is balanced with production cost.

Research Assignment: Design and carry out experiments to determine the quality, quantity, and distribution of mix-ins for a series of commercial ice cream samples.

Instructor Notes for Student Investigation 5B

In this research investigation, students evaluate a number of ice cream samples that contain particles of recognizable size. The amount and quality of mix-ins should be compared for ice creams of different price ranges. Size and shape of nut pieces should be noted and chocolate chip samples should be examined for presence of actual small chocolate chips versus chocolate flakes.

The mass percentage of the particles can be efficiently assessed by placing a known mass of the sample in a sieve and running tepid water over it to wash away the nonsolid portion, then recovering and weighing the solid material. Sampling the top, middle, and bottom of a quart container of ice cream, then determining the number of particles or the mass percentage of mix-ins provides data on distribution.

● Texture of Ice Cream (Student Research Investigation 5C)

The physical structure of ice cream represents a complex mixture of ice crystals, fat globules, colloidal proteins and gums, and air cells dispersed in a liquid phase. Sugars, whey proteins, and soluble salts remain dissolved in the liquid phase. The relationships of these components to one another result in the texture of the ice cream.

An ideal ice cream texture is very smooth, with solid particles too small to be felt in the mouth. If the fat droplets are too large due to incomplete homogenization, they churn easily during freezing and cause the texture to feel greasy or buttery. The most common texture problem is a coarse or icy feel due to ice crystals that are large enough to be detected. When ice cream partially melts and then refreezes, the size of ice crystals grows; therefore, maintaining a steady, low temperature during hardening, storage, and transportation is particularly important. If the ice cream mix contains too much NMSs, some of the lactose may crystallize, causing a grainy (sandy) feel.

When too much air is incorporated into the ice cream during the freezing process, the texture will be snowy or flaky. If not enough air is present, the ice cream will be too dense and difficult to scoop and will resemble a frozen ice cube rather than ice cream.

Research Assignment: Using a homemade ice cream recipe, experiment with various freezing processes to study their effect on texture. Investigate the texture of a variety of commercial frozen products from superpremium ice cream to sherbet.

Instructor Notes for Student Research Investigation 5C

Students should develop a series of experiments to produce ice creams of different textures. One experimental series might study the amount of air incorporated into the mix by freezing some of the mix with no stirring and comparing that to a freezing process by which air is beaten into the mix, continually or at measured intervals. The effect of melting and refreezing on crystal growth is another experimental area easily studied.

Another approach is to have the students make ice cream in a small-capacity home freezer using formulas that have high and low water content. Alternatively, they can prepare a common mix with no gelatin as a control. To this mix they can add 2% additional nonfat dry milk and a package of unflavored gelatin. Varying the amount of cream used is an effective way to change both the water content and the fat content.

Checking differences in texture by the eye is not always easy or accurate. Use a panel of judges to describe differences in texture by marking on a scale the smoothness or iciness of randomly numbered samples. The students can then decode the samples and determine a mean and standard deviation for the samples.

Commercial samples of various types of ice creams should be studied also. Texture of various types of ice cream, from low-fat to superpremium, will be noticeably different, even under hand lens magnification. Sherbets show a markedly more "icy" texture.

● Structure of Ice Cream (Overhead 5)

a: ice crystals

b: air cells

c: unfrozen material

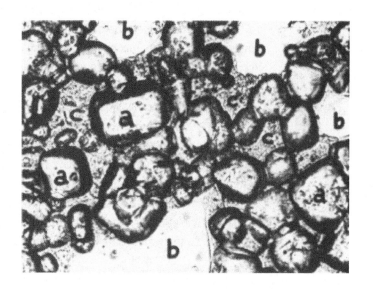

The internal structure of ice cream

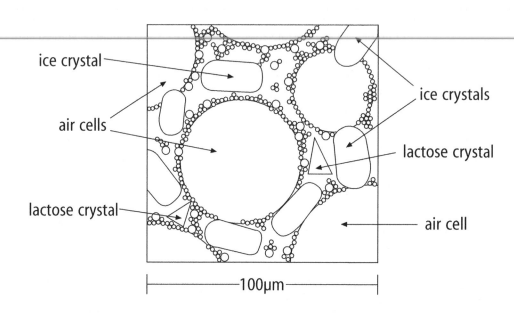

ice crystal

air cells

lactose crystal

ice crystals

lactose crystal

air cell

100μm

Microscopic representation of ice cream structure

12076215R00059